Food Security Challenges and Approaches

Edited by Muhammad Haseeb Ahmad,
Muhammad Imran
and Muhammad Kamran Khan

Published in London, United Kingdom

Food Security Challenges and Approaches
http://dx.doi.org/10.5772/intechopen.100689
Edited by Muhammad Haseeb Ahmad, Muhammad Imran and Muhammad Kamran Khan

Contributors
Christiana V. Abonge, Sujay Kumar Bhajan, Mohammad Mahmood Hasan, Md. Anwarul Haque, Mohammad Nurul Islam, Mahnoor Pervez, Farkhanda Manzoor, Zainab Muhammad Bello, Sanusi Muhammad, Adamu Aliyu Aliero, Ibrahim Aliyu Dabai, Adamou Rabani, Abrar Almalki, Balakrishna Gokaraju, Raymond C. Tesiero, Akinsola Famuwagun, Odunayo C. Adebooye, Tunji V. Odunlade, Kehinde A. Taiwo, Durodoluwa J. Oyedele, Rotimi E. Aluko

First published in London, United Kingdom, 2023 by IntechOpen
IntechOpen is the global imprint of INTECHOPEN LIMITED, registered in England and Wales, registration number: 11086078, 5 Princes Gate Court, London, SW7 2QJ, United Kingdom
Printed in Croatia

British Library Cataloguing-in-Publication Data
A catalogue record for this book is available from the British Library

Additional hard and PDF copies can be obtained from orders@intechopen.com

Food Security Challenges and Approaches
Edited by Muhammad Haseeb Ahmad, Muhammad Imran and Muhammad Kamran Khan
p. cm.

This title is part of the Sustainable Development Book Series, Volume 6
Topic: Sustainable Economy and Fair Society
Series Editor: Usha Iyer-Raniga
Topic Editor: Antonella Petrillo

Print ISBN 978-1-80356-941-3
Online ISBN 978-1-80356-942-0
eBook (PDF) ISBN 978-1-80356-943-7
ISSN 2753-6580

IntechOpen Book Series
Sustainable Development
Volume 6

Aims and Scope of the Series

Transforming our World: the 2030 Agenda for Sustainable Development endorsed by United Nations and 193 Member States, came into effect on Jan 1, 2016, to guide decision making and actions to the year 2030 and beyond. Central to this Agenda are 17 Goals, 169 associated targets and over 230 indicators that are reviewed annually. The vision envisaged in the implementation of the SDGs is centered on the five Ps: People, Planet, Prosperity, Peace and Partnership. This call for renewed focused efforts ensure we have a safe and healthy planet for current and future generations.

This Series focuses on covering research and applied research involving the five Ps through the following topics:

1. Sustainable Economy and Fair Society that relates to SDG 1 on No Poverty, SDG 2 on Zero Hunger, SDG 8 on Decent Work and Economic Growth, SDG 10 on Reduced Inequalities, SDG 12 on Responsible Consumption and Production, and SDG 17 Partnership for the Goals

2. Health and Wellbeing focusing on SDG 3 on Good Health and Wellbeing and SDG 6 on Clean Water and Sanitation

3. Inclusivity and Social Equality involving SDG 4 on Quality Education, SDG 5 on Gender Equality, and SDG 16 on Peace, Justice and Strong Institutions

4. Climate Change and Environmental Sustainability comprising SDG 13 on Climate Action, SDG 14 on Life Below Water, and SDG 15 on Life on Land

5. Urban Planning and Environmental Management embracing SDG 7 on Affordable Clean Energy, SDG 9 on Industry, Innovation and Infrastructure, and SDG 11 on Sustainable Cities and Communities.

The series also seeks to support the use of cross cutting SDGs, as many of the goals listed above, targets and indicators are all interconnected to impact our lives and the decisions we make on a daily basis, making them impossible to tie to a single topic.

Meet the Series Editor

Usha Iyer-Raniga is a professor in the School of Property and Construction Management at RMIT University. Usha co-leads the One Planet Network's Sustainable Buildings and Construction Programme (SBC), a United Nations 10 Year Framework of Programmes on Sustainable Consumption and Production (UN 10FYP SCP) aligned with Sustainable Development Goal 12. The work also directly impacts SDG 11 on Sustainable Cities and Communities. She completed her undergraduate degree as an architect before obtaining her Masters degree from Canada and her Doctorate in Australia. Usha has been a keynote speaker as well as an invited speaker at national and international conferences, seminars and workshops. Her teaching experience includes teaching in Asian countries. She has advised Austrade, APEC, national, state and local governments. She serves as a reviewer and a member of the scientific committee for national and international refereed journals and refereed conferences. She is on the editorial board for refereed journals and has worked on Special Issues. Usha has served and continues to serve on the Boards of several not-for-profit organisations and she has also served as panel judge for a number of awards including the Premiers Sustainability Award in Victoria and the International Green Gown Awards. Usha has published over 100 publications, including research and consulting reports. Her publications cover a wide range of scientific and technical research publications that include edited books, book chapters, refereed journals, refereed conference papers and reports for local, state and federal government clients. She has also produced podcasts for various organisations and participated in media interviews. She has received state, national and international funding worth over USD $25 million. Usha has been awarded the Quarterly Franklin Membership by London Journals Press (UK). Her biography has been included in the Marquis Who's Who in the World® 2018, 2016 (33rd Edition), along with approximately 55,000 of the most accomplished men and women from around the world, including luminaries as U.N. Secretary-General Ban Ki-moon. In 2017, Usha was awarded the Marquis Who's Who Lifetime Achiever Award.

Meet the Volume Editors

Dr. Muhammad Haseeb Ahmad is an Assistant Professor in the Department of Food Science, Government College University Faisalabad, Pakistan. Prior to this, he was an assistant professor at the National Institute of Food Science and Technology, University of Agriculture Faisalabad, Pakistan. He received his doctoral degree from Hohenheim University, Germany, in 2016. During his stay there, he served as a research associate for projects in various food disciplines. Dr. Ahmad is the author of fifty publications, including twenty book chapters. He has presented his work at various national and international conferences. His area of research is food science with a focus on process analytics and data mining.

Dr. Muhammad Imran is an Associate Professor in the Department of Food Science, Faculty of Life Sciences, Government College University Faisalabad, Pakistan. He obtained a Ph.D. in Food Technology from the National Institute of Food Science and Technology, University of Agriculture Faisalabad, Pakistan. He completed an Indigenous Fellowship Program as well as an International Research Support Innovation Program (IRSIP) fellowship. He has expertise in extrusion technology, microencapsulation, lipids chemistry, sensory evaluation, and food process engineering. He has authored more than 100 publications in scientific journals and written 15 book chapters. He has attended several conferences as an invited and keynote speaker, as well as numerous seminars and workshops.

Dr. Muhammad Kamran Khan obtained a Ph.D. from the University of Avignon, France, after which he became an assistant professor in the Department of Food Science, Government College University Faisalabad, Pakistan. He was promoted to associate professor in 2021. He has published numerous research and review articles in scientific journals. The primary focus of his research is the enhancement of technical skills and knowledge in food chemistry with a special focus on polyphenols (bioactive compounds). Dr. Khan received funding from the Higher Education Commission for a research project on fruit processing modes. He completed an Endeavor Research Fellowship at Curtin University, Perth, Australia.

Contents

Preface

Food is a basic necessity and thus everyone should have access to adequate amounts of safe and high-quality food. The concept of food security is of utmost importance for both developed and developing nations. Food availability, access, utilization, and stability are the pillars of food security. Today, these pillars are being negatively affected by natural disasters, poor agricultural and post-harvest practices, climate change, and poor manufacturing and marketing strategies. Food insecurity results in fear of hunger and starvation as well as in various chronic diseases due to reduced immunity, which in turn increases the economic burden on countries worldwide. This book discusses the challenges of food insecurity and their effects on quality of life, the environment, and more. It also discusses alternative means of food production and the development of genetically modified food to combat food security challenges. Finally, the book presents new trends and future prospectives.

Muhammad Haseeb Ahmad, Muhammad Imran and Muhammad Kamran Khan
Department of Food Science,
Government College University Faisalabad (GCUF),
Faisalabad, Pakistan

Section 1

Introduction

Chapter 1

Introductory Chapter: Concept of Food Security and Its Overview

Muhammad Faizan Afzal, Muhammad Haseeb Ahmad, Muhammad Imran, Muhammad Kamran Khan, Muhammad Zubair, Sidra Akram and Muhammad Armghan Khalid

1. Introduction

Food security is a primary concern of significant interest to practitioners, academics, and policymakers all over the globe. Food insecurity is also imperative due to its practical influence on each aspect of social life [1]. Food security is significant as successive food shortages and the food price crisis from 2007 to 2008 brought attention to the substantial importance of food security to ensure political constancy. The importance of food security to public health is incontestable due to the fact that 870 million people throughout the world consume less amount of calories than their requirements, and a plethora of mental and physical health repercussions are associated with such deprivation [2]. Food insecurity assessments and projections determine government policy like Feed the Future Initiative and aid decisions that affect billions of people. Food security helps to maximize economic potential because a malnourished population is less productive. In 2012, Barack Hussein Obama declared that food and nutrition security is an economic concern for the well-being of human beings. Thus, food security is also important from the moral viewpoint that everyone has the equal right to food, clothing, a place to live, and medical care that are good for their health and well-being as described in the Universal Declaration of Human Rights (article 25) [3].

Over the last few decades, the idea of food insecurity has developed and expanded in scope. Initially, the primary emphasis was on the availability and production of food [4]. Subsequently, the scope of the concept was broadened to incorporate not only the availability of food but also its production and availability from various ways like sociocultural, physical, and economic perspectives and the utilization of food. Finally, it was extended to incorporate the stability of these measurements [5].

The term "sustainable development" refers to an approach to economic growth that satisfies the requirements of the current situation without affecting the capacity of the next generations to satisfy their requirements [6]. It has been brought up in the context of international debates in order to introduce the concept of sustainability [7]. Sustainability involves a dynamic nature to prescribe the export-related issues occurred during social, economic, and environmental aspects.

According to FAO, [8] food security states that when everyone has physical and economic access to enough, nutritious, and safe food to suit their food choices and dietary needs for a healthy life. Food security involves various factors like suitable food choices, fair prices, advancement in food safety, accessibility to open as well as competitive markets, healthy diets, and a more sustainable food supply chain [9]. Food security is a combination of various problems interlinked with food utilization, accessibility, and availability [1]. Various other issues like water pollution, economic collapse, currency fluctuations, HIV-AIDS, climate change, trade contracts, and political conflicts affect these factors. Some factors play a pivotal role in causing food insecurity, including poverty, low income, lack of education, high food prices, high unemployment rates, scarce property rights, and partial access to markets [10].

2. Food security terminology

Food security is a multidisciplinary and multisectoral approach; despite its importance, food security has many meanings and operationalization. In fact, a vast range of terminology has been used to debate, measure, and intervene in food security. A wide variety of academic fields like economics, nutrition, agriculture, anthropology, public policy, and sociology, as well as a large number of governmental and nongovernmental as well as national and international agencies, have been actively involved within it. The concept of food security is stated in various contexts, each of which has been accompanied by its particular jargon. Hunger is also associated with food insecurity; the concept of hunger is probably so emotionally intense [8].

The definition of food security is described in the World Food Summit, 1996, which is being commonly used. According to it, food security is accomplished when all people have economic and physical access to a safe, nutritious, and adequate amount of food to accomplish their dietary requirements as well as food choices for a healthy life [11]. According to the definition, food insecurity occurs when one or more of the above conditions are not met. Food insecurity is categorized into three types: chronic, seasonal, and temporary food insecurity. The most severe form is chronic food insecurity, whereas transitory food insecurity is the least severe [12].

People sometimes use the terms "food security" and "nutrition security" synonymously, but the definitions of each are much different. Food security is necessary for nutrition security, but it is not enough. Aside from food, nutrition security also looks at care, health, and hygiene practices. According to FAO, nutrition security is stated as when everyone in a household has easy access to a healthy, well-balanced diet, a clean environment, and good health services and care. This makes sure that everyone is healthy and able to live a good life [2].

3. Evolution concept of food security

The notion of food security started about 45 years ago when there were food shortages all over the world. At first, it focused on making sure that there was enough food and that the prices of basic foods were stable around the world and in each country. This happened because agricultural commodity prices were very unstable in the early 1970s. This was caused by a disarray in the monetary systems and financial marketplaces as well as various other unfavorable conditions. The attention to supply constraints was a result of the changing structure of the global food economy, which is thought to be the cause of these economic problems. Hunger, food crises, and famine emphasized to develop the definition of food security that considers the most important needs and actions of people who might be affected [13, 14].

In 1994, UNDP Human Development Report examined human security in which seven key threats were recognized including food, economic, environmental, communal, political, personal, and health security [15]. Human rights entered the concerns about food security around this time [16].

Food insecurity is a problem that ranges from the household to the international level, despite the fact that it is primarily addressed at the national level by policymakers. At the household level, the measurements of food security occur in order to take into account individual dietary preferences; thus, it is considered that food insecurity is a problem that affects the whole world. By 1993, food security had nearly 200 definitions [17]. This unpredictable condition showed that food security studies depended on the technical opinions, context-specific and policy issues being considered.

The subsequent momentous occasion occurred when the World Bank published the foundational report entitled "Poverty and Hunger" in 1986 [18]. A difference between chronic food insecurity and temporary as well as acute food insecurity, which is associated with poverty and human-made or natural calamities, was observed, respectively. This resulted in the introduction of a temporal scale for measuring food security. This was considered when expanding the idea of food security to include the following parameter: availability of sufficient, safe, and nutritious food to people at all times for a dynamic and healthy life [19]. The second edition of the notion took place in 1994, in response to the Human Development Report issued by the UN Development Program, which examined the prerequisites for human food security. In this situation, food security, which was part of the overarching concept of social security, involved the conversion of human rights.

After the World Food Summit (1996), a very well-organized and systematic work was endorsed in which the organization's members tasked it with monitoring the progress toward decreasing the number of malnourished people in poor countries by 2015 [20]. In the middle of 1990, the term food security was replaced by nutrition security and food and nutrition security to elaborate on the current scenario [21]. The annual flagship report of FAO, entitled The State of Food Insecurity in the World, was published in 1998 and used to evaluate the whole monitoring process.

The statement of food security was improved by The State of Food Insecurity in the World 2001; according to this statement, food security is a discipline ensuring that all the people at the same place have equal rights to sufficient, hygienic, nutritious, adequate, and safe food to fulfill their dietary needs as well as food choices with physical and economic access for a healthy life. Furthermore, improving poverty is mandatory in addressing food security but is insufficient [22, 23].

Various topics like investments in agriculture, social safety, food losses and waste, land tenure, biofuels, and price instability impacted the food security recognized by the Committee on World Food Security (CFS) in 2009 reforms. These topics were discussed in the sustainable concerns about fisheries, aquaculture, and food systems based on the High-Level Panel of Experts on Food Security and Nutrition (HLPE) discussion. At last, in 2014 and 2015, HLPE reports requested by CFS revealed progressive incorporation of sustainability into food and nutrition security [24].

It was acknowledged that eliminating poverty is crucial for achieving this goal but that doing so will not be adequate on its own [23]. During the World Summit on Food Security in 2009, the concept of food security received its most recent formal revision, which included the addition of the fourth dimension of stability [5]. In more recent times, a fifth dimension has been also proposed, which would be sustainability, to be added in order to integrate the long-term sequential aspect [19].

4. Food security and its dimensions

According to the definition, there have been identified four different aspects of food security [12]. Availability of food is the first dimension of food security that deals with the production of food locally and bringing it in from other countries. Second is the accessibility of food to everybody. The consumer is able to pay for the food, and the food can reach the consumer.

Along with such financial and physical availability, sociocultural availability assures that the food is appropriate for the culture, and social safety networks are available to help those who are less fortunate. Utilization is the third dimension that describes that a person realizes his extraordinary potential and lives a successful as well as healthy life; they have sufficient amounts of food both in terms of quantity and quality. The ability of a nation, community, and a person to absorb the shocks of the system of the food chain, whether those stresses are generated by natural calamities like climate change and earthquakes or those that are man-made, also underlies the domain of stability, which describes specific topics like wars and economic crisis. Sustainability indicators at a global or local level include biodiversity, climate change, and ecology in addition to socioeconomic and sociocultural variables [19].

5. Food security and sustainability

Sustainability is a generic notion applicable to food security as a whole, representing more than its particular dimensions. The idea of sustainability emphasizes that continuous healthy well-being across the lifetime and future generations is the objective of long-term food security [25]. Food insecurity can happen whenever there is an interruption at any step along the pathway from availability to use and consideration to stability because the four dimensions are linked and interdependent. However, the most significant complication and tragedy arise from these considerations. As can be seen from the definitions of each term, the concept of "sustainable diets" is an attempt to bridge the gap between "sustainability" and "food security" in order to create "holistic, sustainable food systems." Diets are considered sustainable if they "have preventive aspects, reverential of biodiversity and ecosystems, socially desirable, available, economically sustainable, low-cost; nutritionally adequate, hygienic, and healthy; and enhancing human as well as natural resources" [23]. Whereas "a sustainable food system ensures Food Security and Nutrition for everyone in that way which does not hurt the social, environmental and economic bases for ensuring FSN for future generations" [26].

The widespread consensus among nations worldwide is that climate change poses a risk to the permanent viability of food security. However, the processes included in food systems are responsible for approximately (20–30%) all greenhouse gas emissions related to human activity, and as a result, climate change occurs [27]. It was observed that a relationship that might be described as a "trade-off" exists between reducing human-caused greenhouse gas emissions and ensuring food security within the context of the food system that is already in place. As a result, a systematic and comprehensive method is required in order to satisfy the immediate and long lasting needs of food security while also reducing the adverse effects on the environment caused by greenhouse gas emissions caused by the activities involved in the food system itself. Our understanding is always expanding, despite the fact that it is not entirely obvious what sustainable food systems might look like in practice [28].

6. Concept of food insecurity

Food insecurity will happen when there are problems at any one level of the food production-consumption pathway. The FINS level upstream has a big effect on those downstream. Food insecurity is when adequate, safe, and nutritious foods are unavailable or when there is uncertainty about how to get acceptable foods in a socially acceptable way [29]. In the United States, food insecurity occurs when the access and availability of food for the future are uncertain, not enough food for a healthy lifestyle, or there is the need to adopt unethical methods to obtain food [30]. Food insecurity can occur for several reasons, the most common of which is a lack of financial resources; however, it can also occur when food is readily available and easy to obtain, but it is unable to be consumed due to physical limitations, like the partial physical functioning of the elderly and disabled individuals [30]. Despite this, given the current emphasis on health fairness, priority should be given to individuals living in the most precarious conditions. They are being put under pressure by a variety of calamities, both natural and artificial, like droughts, wars, floods, and conflicts. There is an imperative need for improved methods of managing the precarious state of food supply. In a perplexing turn of events, the populations with the highest levels of food insecurity, like immigrants, the homeless, and displaced persons, are typically excluded from food security surveys, which leads to an underestimation of the extent of the issue.

7. Measuring food security

7.1 Food security at national level estimates

Food security measures that are meant to be used at a national level often focus on how easy it is to get food. Food balance sheets, which are used to figure out how much food is available, have traditionally been based on national data on the total amount of food produced and imported (food supply) and the amount of food that is exported for many reasons, such as to be used as seed, to feed livestock, to process food and nonfood, or to be lost during storage and transportation [31]. The FAO's main way to measure food security is by the number of people who are undernourished. These data are used to make that number.

Food supply and usage data are valuable for anticipating food shortages and surpluses, projecting future food demand, and setting agricultural output targets [31]. It was observed that the average dietary energy intake was equal to the mean value of calorie consumption of the whole population [2]. This is an unjustified and inconsistent assumption. Despite the deficiency of comprehensive information on food distribution balance and losses, considerable differences were detected between these data and USDA estimates of food-insecure families [32]. The USDA uses income distribution statistics and aggregated food supply estimates to project the consumption of calories for different income groups [33].

7.2 Global food security index

Food Security Index at the global level is another way to measure food security on a national level. It is based on a number of different factors. It was thought up and made by the Economist Intelligence Unit, which is one of the many parts of the Economist Group, a publicly traded multinational company. The project was paid for by the company DuPont. A total of 30 different variables were used to provide a standard for comparing the food security of different countries. These indicators are split into three groups: availability, affordability, and quality and safety, each of which has 10, 6, and 14 indicators [34].

The Global Food Security Index assesses the success of countries in accomplishing food security, much like other national-level metrics; however, it does so by utilizing quantitative and qualitative measures that represent not only food availability but also food access (for example, food intake as a proportion of total household expenditure and proportion of population living below or close to the global poverty line, food costs) and diet quality such as dietary availability of micronutrients. The GFSI is adjusted every quarter to account for changes in the data about food prices.

The Global Food Security Index is computed on a regular schedule in response to changes in the data on food prices. The GFSI uses data from the World Bank, the Food and Agriculture Organization of the United Nations, the World Food Program, the Economist Intelligence Unit, analysts and expert panels, and public as well as academic sectors. These experts give the subjective scores that are used to make many of the qualitative indicators that are used to make the index. They also decide which indicators will be in the index and how much weight each one will have. The reliance on consensus and expert opinion is different from the methods used by FAO and IFPRI. However, personal data interpretation is often used when making food security measures, as explained below. Because of the complexity of food security indicators and the need for context to understand them, some institutions focus on consultative methods when making the tools for measuring food security.

FAO, the World Food Program, and the International Fund for Agricultural Development have come up with signs for food security in the past few years [23]. Several indicators describe each aspect of food security. A lot of work is also being done to combine these indicators into indices. These were chosen from many different indicators based on how important they were, how easy they were to get, and how often they were measured.

7.3 Vulnerability analysis and mapping methodology

When conducting analyses on food security, the WFP makes use of various types of evaluations, collectively referred to as vulnerability investigation and mapping. The CFSVAs stand out as the most important of these evaluations. These studies are conducted in nations that are prone to crises and suffer from food insecurity in order to evaluate the current state of food security and investigate the underlying causes of susceptibility [35].

They obtain primary data by surveying households and rely on secondary data analytics for their research. These surveys contain 13 basic food security evaluation systems, which exemplify the complexity of measuring food security. These modules include food consumption patterns, water sources, sanitation access, household assets, household composition, housing materials, sustainable income livelihood, education, credit availability, livestock, agriculture, external aid, shocks, and coping mechanisms [35]. WFP conducts a variety of methods like food security monitoring ways, crop availability assessment operations, market assessment, emergency food security evaluation, and combined assessment operations as part of its goal to increase governments' ability to decrease hunger [36].

7.4 Household measuring food access

Some of the more talked-about ways to measure food security only describe the national food supplies. But because they focus on estimates and trends at the national or regional level, they do not pay much attention to how households act and other factors that affect food access. At the household level, there are a number of tools that can be used to measure changes in food security between and within households. Because these measures use information from household surveys, they can measure the "access" part of food security more accurately than those measures that use information from the whole country. But various methods used to measure food access actually measure how people get food or how much food they eat. Food access means being able to get food both physically and financially. However, numerous tools are used to measure food access to measure how much food is bought. It is a standard practice to use both terms interchangeably when referring to food access; nonetheless, it is vital to differentiate between the two for the sake of measurement [37].

7.5 Food security and dietary adequacy

The evaluation of a nation's ability to ensure its own food supply has expanded beyond the simple assessment of the quantity of food available to encompass other evaluations of economic access. On the other hand, the same level of focus has not been placed on evaluating the quality of diets. If the data from HCESs are collected correctly, it may be possible to estimate the quality of the diets of a whole population. Some of the data collected to help make CSIs may also tell us about the quality of people's diets. Most food security assessments do not use more severe measures of diet quality like 24-h dietary recall and food questionnaires because it takes a lot of time and money to collect and analyze these kinds of data.

Another component of food availability, that is, the embarrassment that comes along with acquiring food in methods glared upon by society, is noticeably inattentive from the metrics examined here, in spite of the fact that social unacceptability has been established as a shared domain of food insecurity across a wide variety of cultural contexts [38, 39]. In point of fact, a query investigating it was included in the early development stages of the HFIAS; however, it was removed from the completion of the measure due to the delicate nature of the subject matter and the trouble in obtaining precise reports [40]. Limited surveys inquire about food acquisition acceptability; hence insignificant data is available to examine this food security section.

Food safety is generally excluded from food security metrics despite growing issues about mycotoxin contamination, chemical pollutants, foodborne illness, and zoonotic disease [41].

7.6 Categorization of food insecurity

Determining acceptable parameters to employ both within a setting and when comparing results from other areas or countries is another challenge that must be overcome when attempting to quantify food security. Several different food security indicators clarify the plan of limitations for determining levels of food insecurity, which provide precise information on the topic.

The IPC provides precise classification standards since this classification directly defines the priority response objectives it strives to achieve. Food insecurity is classified into categorical groups that are not as significant for other measures, such as the CSI, FCS, and HFIAS, which provide quantitative scores. Instead, comparing values' fluctuation across a specific population may be more helpful. It was observed that the persistent and present food insecurity complications are associated with severe and moderate food insecurity parameters. A variety of food insecurity conditions like chronic, moderate chronic, severe transitory, and transitory food insecurity are briefly described by Devereux [42]. It is essential to accomplish the limits, and the classification of food insecurity is described well, regardless of the approach used for measuring and categorizing food insecurity.

7.7 Future agenda for food security

In the next century, to measure global food security, we need to evaluate the effect of climate change in the food production capacities. A broader definition of food security increases the bar for success and gives numerous paths to achieve it. We should also talk about how income is shared, what people eat, and how much food is lost or wasted. Better education about health and nutrition will help people, families, and communities get more food. By making emergency response systems and other social safety nets stronger, we can reduce food insecurity and make society more resilient as a whole.

Overall, food delivery must expand to accommodate population expansion and improving diets. Long-term growth must match the past 50 years if the population doubles in 66 years. Food production efficiency could contribute to this growth. Ongoing discussions over the sustainability of historic growth rates bring such continuous growth into question, especially in light of climate change. Without this growth, achieving global food security would be challenging. Even with the more extensive food availability and access and projected rises of 3 to 4 times in food production and four to six times in income, ensuring food security will need consolidating the human right to food, a much enlarged safety net, and an improving ability to adjust to surprise. Social institutions, value changes, and increasing flexibility already exist.

8. Future prospects

If the amount of food that can be produced in the next 40 years needs to increase by 50 percent despite the amount of available land decreasing, then this will require a massive and ongoing investment of money, struggle, and time. Similar to the historical achievements of global agriculture, which resulted in the "green revolution" and prevented millions of people from dying of starvation, a significant portion of the solution will need to originate from technological advancements. Thus, this advanced technology will aim to ensure safe, sufficient, and nutritious food to fulfill the dietary needs of the worldwide human population. Such technologies are beneficial to produce, distribute, and process food at the market level for easy access of consumers. Therefore, the investment in the development of agriculture and long-term neglected worldwide research should be double for the proper application of development of these technologies.

Any debate of food security is insufficient without addressing the trade and complicated network of sociopolitical and other challenges, which are more precise as compared to processing and production. The effects of plant diseases on food production, quality, and safety were mainly mitigated by climate change. The overview of this chapter emphasizes plant protection professionals share their scientific knowledge to inform policy debates. To boost productivity and quality, research must be communicated beyond the farm gate to policymakers and the public.

There has been limited experimental observation on plant diseases under field conditions that accurately imitate climate change, limiting opportunities to improve crop adaptability or disease management. In addition, through the use of models, a significant amount of information regarding the potential effects of global climate change has been acquired. Initial assessments are available for a select number of nations, regions, crops, and diseases. Thus, food security is a necessary measure to ensure safe, sufficient, and nutritious food for human beings all over the world.

Many climate change worrywarts assume a changing climate would make food security more challenging. If climate changes occur, they could affect food production and revenue. Climate change mitigation or adaptation may slow output growth or redirect funds from poverty or food insecurity relief. Adapting to climate change and making sure there is enough food can both help each other. As places build up their agricultural infrastructure over the next few decades, they may be able to take advantage of their great growing conditions. Second, the fact that climate change affects the whole world encourages everyone to do their part and work together. During talks between rich and poor countries about possible preventative and adaptive actions, food security could be used as an example of a low-cost goal with a high return in terms of building international cooperation and creative skills. Food security for industrialized and developing countries will require proactive policies, programs, and institutions.

9. Conclusion

The importance of food security to public health is incontestable due to the fact that 870 million people throughout the world consume a smaller number of calories than their requirements. There is an imperative need for improved methods of managing the precarious state of food supply. Food access refers to economic as well as physical access to food; though numerous tools are used to measure food access to measure food procurement, food safety is generally excluded from food security metrics despite growing issues about mycotoxin contamination. Many climate change worrywarts assume a changing climate would make food security more challenging.

Author details

Muhammad Faizan Afzal[1], Muhammad Haseeb Ahmad[1*], Muhammad Imran[1], Muhammad Kamran Khan[1], Muhammad Zubair[2], Sidra Akram[2] and Muhammad Armghan Khalid[1]

1 Department of Food Science, Government College University Faisalabad, Faisalabad, Pakistan

2 Department of Home Economics, Government College University Faisalabad, Faisalabad, Pakistan

*Address all correspondence to: haseeb.ahmad@gcuf.edu.pk

References

[1] Martin-Shields CP, Stojetz W. Food security and conflict: Empirical challenges and future opportunities for research and policy-making on food security and conflict. World Development. 2019;**119**:150-164

[2] FAO. World Food Programme, IFAD. The State of Food Insecurity in the World 2012. Rome: FAO; 2012

[3] United Nations. Universal Declaration of Human Rights. Paris: United Nations; 1948

[4] United Nations. Report of the World Food Conference, Rome, 5-16 November 1974. New York: UN; 1975

[5] Food and Agriculture Organization of the United Nations. Declaration of the World Food Summit on Food Security. Rome: FAO; 2009

[6] Holden E, Linnerud K, Banister D. Sustainable development: Our common future revisited. Global Environmental Change. 2014;**26**:130-139

[7] United Nations. Our Common Future: Report of the World Commission on Environment and Development. Geneva: UN; 1987

[8] Mason JB. Measuring Hunger and Malnutrition: Methods for Measuring Food Deprivation and Undernutrition. Rome: FAO; 2002

[9] Armghan KM, Niaz B, Saeed F, Afzaal M, Islam F, Hussain M, et al. Edible coatings for enhancing safety and quality attributes of fresh produce: A comprehensive review. International Journal of Food Properties. 2022;**25**(1):1817-1847

[10] Lampietti JA, Michaels S, Magnan N, McCalla AF, Saade M, Khouri N. A strategic framework for improving food security in Arab countries. Food Security. 2011;**3**(1):7-22

[11] FAO. Rome Declaration on World Food Security and World Food Summit Plan of Action. Rome: FAO; 1996

[12] FAO. Food Security Information for Action Practical Guides. Rome: FAO; 2008

[13] Clay E. FAO Expert Consultation on Trade and Food Security: Conceptualizing the Linkages. Rome: FAO; 2002

[14] Shaw DJ. World Food Security. A History since 1945. New York: Palgrave Macmillan; 2007

[15] United Nations Development Programme. Human Development Report. Oxford and New York: Oxford University Press; 1994

[16] Drèze J, Sen A. Hunger and Public Action. Oxford: Clarendon Press; 1989

[17] Smith M, Ponting J, Maxwell S, et al. Household Food Security, Concepts and Definitions: An Annotated Bibliography. Brighton: Institute of Development Studies; 1993

[18] World Bank. Poverty and Hunger: Issues and Options for Food Security in Developing Countries. Washington, DC: World Bank; 1986

[19] Berry EM, Bernini S, Burlingame B, Meybeck A, Conforti P. Food security and sustainability: Can one exist without the other? Public Health Nutrition. 2015;**18**:2293-2302

[20] McDonald BL. Food Security. Polity; 2010

[21] Committee on World Food Security. Coming to Terms with Terminology. Rome: FAO; 2012

[22] Food and Agriculture Organization of the United Nations. The State of Food Insecurity in the World 2001. Rome: FAO; 2002

[23] Food and Agriculture Organization of the United Nations, World Food Programme & International Fund for Agricultural Development. The State of Food Insecurity in the World. Economic Growth Is Necessary but Not Sufficient to Accelerate Reduction of Hunger and Malnutrition. Rome: FAO; 2012. p. 2012

[24] Committee on World Food Security. 35th Session, Agenda Item III, Reform of the Committee on World Food Security. Rome: FAO; 2009

[25] Berry EM, Degeest S. Tell me what you eat and I will tell you your sociotype: Coping with diabesity. Rambam Maimonides Med. 2012;**3**:e0010

[26] HLPE. Nutrition and Food Systems: HLPE Report 12. A Report by High Level Panel of Experts on Food Security and Nutrition of the Committee on World Food Security (HLPE). Rome; 2017

[27] Garnett T, Smith P, Nicholson W, Finch J. Food Systems and Greenhouse Gas Emissions (Food Source: Chapters). Food Climate Research Network: University of Oxford; 2016a

[28] Garnett T, Smith P, Nicholson W, Finch J. Overview of Food System Challenges (Food Source: Chapters). Food Climate Research Network: University of Oxford; 2016b

[29] Clapp J, Moseley WG, Burlingame B, Termine P. The case for a six-dimensional food security framework. Food Policy. 2022;**106**:102164

[30] National Research Council. Food Insecurity and Hunger in the United States: An Assessment of the Measure. Washington, DC: National Research Council; 2006

[31] FAO, WFP and IFAD. The State of Food Insecurity in the World 2013: The Multiple Dimensions of Food Security. Rome: FAO; 2013

[32] FAO. Food Balance Sheets: A Handbook. Rome: FAO; 2001

[33] Barrett CB. Measuring food insecurity. Science. 2010;**327**:825-828

[34] Shapouri S, Rosen S, Meade B, Gale F. Food Security Assessment, 2008-9 Outlook GFA-20. Washington, DC: USDA Economic Research Service; 2009

[35] Economist Intelligence Unit. Global Food Security Index Examines the Core Issues of Food Affordability, Availability and Quality [Press Release]. London: Economist Intelligence Unit; July 2012

[36] World Food Programme. Comprehensive Food Security and Vulnerability Analysis Guidelines. 1st ed. Rome: World Food Programme; 2009

[37] Jones AD, Ngure FM, Pelto G, Young SL. What are we assessing when we measure food security? A compendium and review of current metrics. Advances in Nutrition. 2013;**4**(5):481-505

[38] World Food Programme [Internet]. 2013. Available from: http://www.wfp.org/food-security/assessments. [Accessed: January 14, 2013]

[39] Coates J, Wilde PE, Webb P, Rogers BL, Houser R. Comparison of a qualitative and a quantitative approach to developing a household food insecurity scale for Bangladesh. The Journal of Nutrition. 2006;**136**:S1420-S1430

[40] Radimer KL, Olson CM, Greene JC, Campbell CC, Habicht JP. Understanding hunger and developing indicators to assess it in women and children. Journal of Nutrition Education. 1992;**24**:S36-S44

[41] Coates J, Swindale A, Bilinsky P. Household Food Insecurity Access Scale (HFIAS) for Measurement of Food Access: indicator Guide. Washington, DC: Academy for Educational Development; 2006

[42] WHO. Guidelines for Developing and Implementing a National Food Safety Policy and Strategic Plan. Brazzaville: WHO Regional Office for Africa; 2012

Section 2

Sustainable Approaches for Food Security

Chapter 2

Potential of Indigenous Vegetable-Fortified Food Products for Improved Human Nutrition and Health in West Africa

Albert A. Famuwagun, Odunayo C. Adebooye, Tunji V. Odunlade, Kehinde A. Taiwo, Durodoluwa J. Oyedele and Rotimi E. Aluko

Abstract

The study developed and tested the acceptability of key innovative value-added products, including vegetable-fortified bread and pastry products (cookies and chinchin), using each of these three indigenous vegetables, namely *Telfairia occidentalis* Hook. F. (fluted pumpkin), *Amaranthus virdis* L. (local amaranth), and *Solanum macrocarpon* L. (eggplant). The bread was fortified with each of the individual dried leafy vegetables at levels of 1%, 3%, and 5%. Ogi was also fortified with the equivalent of the dried *Telfaria occidentalis* in the wet form at 1%, 3%, and 5%, while pastry was fortified with the only extract obtained from amaranth leafy vegetable at 1%, 3%, and 5%. Results from the MicroVeg project showed that bread fortified with 3% (w/w) dried vegetable leaf powders had higher nutritional properties. Pastry formulations that included 3% dried vegetable leaf powders had slightly lower consumer acceptability in terms of taste and color; however, due to the associated potential health benefits, consumers were willing to buy the vegetable-fortified products as replacements for the regular products. This chapter discusses the full details of the nutrition and value addition aspects of MicroVeg with empirical examples of the interventions and the potential contributions to dietary diversity and enhanced family nutrition.

Keywords: MicroVeg project, indigenous vegetables, fortification, baked foods, maize porridge

1. Introduction

As previously reported by World Health Organization [1–3], a low intake of vegetables puts people at the risk for micronutrient deficiencies. Appropriate intakes of fruits and vegetables are known to reduce the risk of cardiovascular diseases and cancer, which are the two most common causes of premature death worldwide and that accounted collectively for 25.5 million deaths in 2013 [4]. The WHO (http://www.who.int/dietphysicalactivity/fruit/en/index2.html) also reports that approximately 14% of gastrointestinal cancer deaths, 11% of ischemic heart disease deaths, and 9% of stroke-related deaths, globally, are due to insufficient intakes of fruits and vegetables. It is also estimated that up to 2.7 million lives could potentially be saved each year if vegetable consumption was sufficiently increased. Given the overwhelming evidence supporting the role of vegetables in the sustenance of human health and well-being, the WHO and Food and Agricultural Organization of the United Nations recommend a minimum intake of 400 g per day of green vegetables [1, 5]. Therefore, to promote and sustain the consumption of vegetables in various forms, a system of including vegetables in the daily human diet is desirable, in form of food fortification to enhance dietary diversity. Food fortification using vegetable products, especially in dessert foods, juices, drinks, and daily flour-based products will constitute a major strategy for making vegetables available for a wide spectrum of the human population.

Over the past 80 years, food fortification, which is a cost-effective and sustainable innovation, has played a vital role in the amelioration of several ailments worldwide [6, 7] because it is a means of reducing the effects of deficiencies that are linked to micronutrients and some macronutrients. The FAO/WHO [8] defined food fortification as the addition of one or more essential nutrients to a food, whether or not it is normally contained in the food, for the purpose of preventing or correcting a demonstrated deficiency of one or more nutrients in the population or specific population groups. Indeed, food fortification has been identified as one of the top four strategies for decreasing micronutrient malnutrition at the global level [9]. Based on the 2017 data on the state of food insecurity in the world, an estimated 2 billion people suffer from hidden hunger around the globe and over 25% of people living in sub-Saharan Africa suffer from chronic hunger. These critical nutritional challenges may be ameliorated through the fortification of staple foods with appropriate nutrients.

Food fortification ensures dietary diversity, improved health, and nutrition, as well as financial returns for small and big businesses [6]. As promising as food fortification sounds, especially in relation to the sustenance of human health, there have been a number of important lessons learned in the research on food fortification processes [8] For example, the development of new combinations of fortificants and vehicles are cost-intensive and time-consuming, while the fortificants must meet quality criteria specifications including, chemical stability, appearance, bioavailability, and homogeneity. Economic analysis indicated that food fortification is a high-priority investment [10]. However, if traditional food vehicles are used to convey these beneficial nutrients, there is potential for cost-saving when considered along with the long-term impact. A major step in popularizing fortified foods is promotion and awareness creation to test the acceptability of the foods.

The International Development Research Centre (IDRC) and Global Affairs Canada (GAC) under the Canadian International Food Security Research Fund (CIFSRF) recently provided research grants, which enabled activities of the indigenous vegetables project (MicroVeg). The MicroVeg project developed innovative systems of production, utilization, and value addition options to enhance nutrition and economic empowerment of the resource poor rural populations in Nairobi and Benin Republic, mainly women. Specifically, the value addition and food fortification studies focused on three indigenous vegetables, which are commonly consumed and of high prominence in both the food chain and markets of the West Africa sub-region: local amaranth (*Amaranthus viridis*), fluted pumpkin (*Telfairia occidentalis*) and African eggplant (*Solanum macrocarpon*). This approach was taken in order to encourage vegetable consumption by using popular foods as a vehicle. The aim of this paper is to describe the food fortification approach used by the MicroVeg project, with emphasis on potential health benefits to the local population, especially women.

2. Key messages

- Vegetable leaf fortification led to the increased nutritional value of maize porridge (*ogi*), a popular weaning food.
- Wheat bread fortified with vegetable leaves (*green bread*) had lower moisture content and higher polyphenol content than unfortified white bread; the increased polyphenol content led to stronger radical scavenging and metal chelating abilities.
- Economic analyses indicate profitability of the vegetable-fortified bread and pastries, with 3% incorporations as the most promising.
- The MicroVeg project created awareness on the food value and affordable/simple value addition technologies through daily radio jingles and direct training of 5466 persons on different aspects of family nutrition and dietary diversity.

2.1 Strategy for value addition studies and scale up

The value addition aspect of the work was conducted in the southwestern part of the country. The major aim of the value addition studies was to scale up the various innovations developed into adaptable forms for the poor resource populace. In achieving this aim, the project involved some Nigerian-based non-profit organizations, such as the Green Generation Initiative and the Institute for Agricultural Research and Training. Through these organizations, the project was able to scale up the innovations and delivered them to the rural end-users using an extension system called the innovation platform (IP) model (**Box 1**), which is aimed at ensuring innovation delivery systems

- Working with multiple actors for consensus building in vegetable value addition.
- Coordination and collaboration along the value addition chain, which is expected to result in more
- Efficient and equitable linkages of the actors in the vegetable value chain.

The IP is based on the thesis that improved interactions, through dialog along the value chain could help to forge linkages among stakeholders, which could result in enhanced communication and information exchange to address common challenges. This propels the adoption of innovation.

Box 1.
Innovation platform (IP) approach has the following activities:

- Identification of target communities and women groups for training on value addition (41 communities).
- Identification of the other actors/stakeholders involved in the vegetable value chain.
- Development of value addition training materials in the local language (Yoruba).
- Series of visits to the communities and women groups for awareness creation.
- Training of project technical staff on value addition innovations and scaling up activities (12 technical officers).
- Launching of the innovation platform at the selected locations.
- Training of 5466 persons (68% women) at 25 locations in value addition innovations.
- Training 100 bakery owners in southwest Nigeria on the production of vegetable-fortified bread.

Box 2.
Steps for establishing an innovation platform

1. Training of the technical officers on the sensory evaluation template (12 staff).
2. Identification the respondents (41 locations, 250 persons/location, total 10,250 panelists).
3. Sensory evaluation implementation.
4. Reporting.

Box 3.
Sensory evaluation of vegetable-fortified foods

that quickly bring innovations to women and rural farmers, therefore making increased adoption possible, notably through appropriate use of new information and communication technologies (**Box 2**). The recipes for the different formulations of vegetable value-added products were designed by the Department of Food Science and Technology, Obafemi Awolowo University, Ile-Ife, Nigeria. The University of Manitoba took the lead and utilized its expertise in food chemistry, food processing food product development, and food/nutritional quality evaluation to enhance product and nutritional quality evaluation of vegetable-fortified products. The University of Manitoba with state-of-the-art facilities was used to train food scientists who were involved in the sensory studies of fortified food products in Nigeria (**Box 3**).

2.2 Fortification of local foods with vegetables

Most popular traditional foods, especially the starch-based local foods in Nigeria and the Republic of Benin are not good suppliers of critical micronutrients, thus establishing the need for food fortification, especially for foods targeted at growing children and vulnerable populations. Newborn babies, especially in the rural areas (65% of the population in Nigeria) are fed with a popular maize-based weaning food, called "*ogi*". In addition, some of the most cherished snacks and desserts, eaten by children and adults, were considered for fortification. Thus, our project developed and tested key innovative value-added products, including vegetable leaf-fortified bread and popular local pastry products (cookies and *Chinchin*). The MicroVeg project conducted a series of studies aimed at improving these traditional foods with each of the three leafy vegetables (amaranth, fluted pumpkin, or African eggplant), including the sensory evaluation and acceptability studies. Green leafy vegetables were chosen as a fortificant because of the abundant bioactive components, polyphenols, that have the potential to defend the human body against degenerative diseases [11, 12]. Ongoing studies by the MicroVeg team have generated results that suggest food fortification with these vegetables could have a significant effect on lowering of blood pressure [13], as well as increasing nutritional and antioxidative properties of the fortified foods [14, 15].

2.3 Fortification of *Ogi*, a local weaning food

Our first case study is on the fortification of a popular Nigerian weaning food known as *ogi* (maize porridge) with a polyphenol-rich extract from leafy vegetables. This study is justified by the state of malnutrition among the women, nursing mothers, and children, in rural Nigeria where fortification of weaning as well as convalescent foods became necessary. For example, UNICEF [16] estimated that 37% of children in Nigeria are stunted (chronically malnourished or low height for age) with more than half of them severely stunted. The same report also highlighted that 18% of children suffer from wasting (acutely malnourished or low weight for height) and 20% of children are underweight (both acutely and chronically malnourished or low weight for age). Several works have been done on the fortification of weaning foods with plant products, such as cocoa solids [17], moringa leaf powder [18], moringa seed [19], pawpaw [20], cowpea [21], melon [22], okra seed meal [23], soybean [24], African yam bean [25], carrot [26] and many others. However, the fortification of maize *ogi* with vegetable polyphenol-rich extract is a promising innovation. The MicroVeg-optimized procedure for the production of vegetable leaf-fortified *ogi* is simple and cost-effective. The maize *ogi* was prepared as a slurry and later dried to produce a cake as shown in **Figure 1**. The vegetable polyphenol was extracted by squeezing vegetable leaves in water at the ratio of 1:2. The extract was filtered to remove vegetable particles and the filtrate was stored in the fridge at 4°C. The *ogi* cake was first made into a slurry by dispersing the cake in water. Thereafter, the polyphenolic filtrate and *ogi* slurry were mixed together at varying proportions (10/100, 25/100, 50/100, 75/100, and 100/100 ml/g of filtrate to slurry) prior to drying at 55°C in a hot air cabinet oven to produce vegetable leaf-fortified *ogi* flour [27]. The vegetable leaf-fortified maize *ogi* powder was reconstituted by adding 1.5 parts of water to make a suspension. To the suspension, 100 ml of hot water was added with continuous stirring to produce maize *ogi* hot porridge of desired consistency. The fortified *ogi* samples were subjected to sensory evaluation, in terms of acceptability of taste, color, flavor, texture, appearance, and overall acceptability.

2.4 Fortification of bread with indigenous vegetables

The development of enriched bakery products and pastries has recently increased and is attracting much attention from researchers, especially in the optimization of

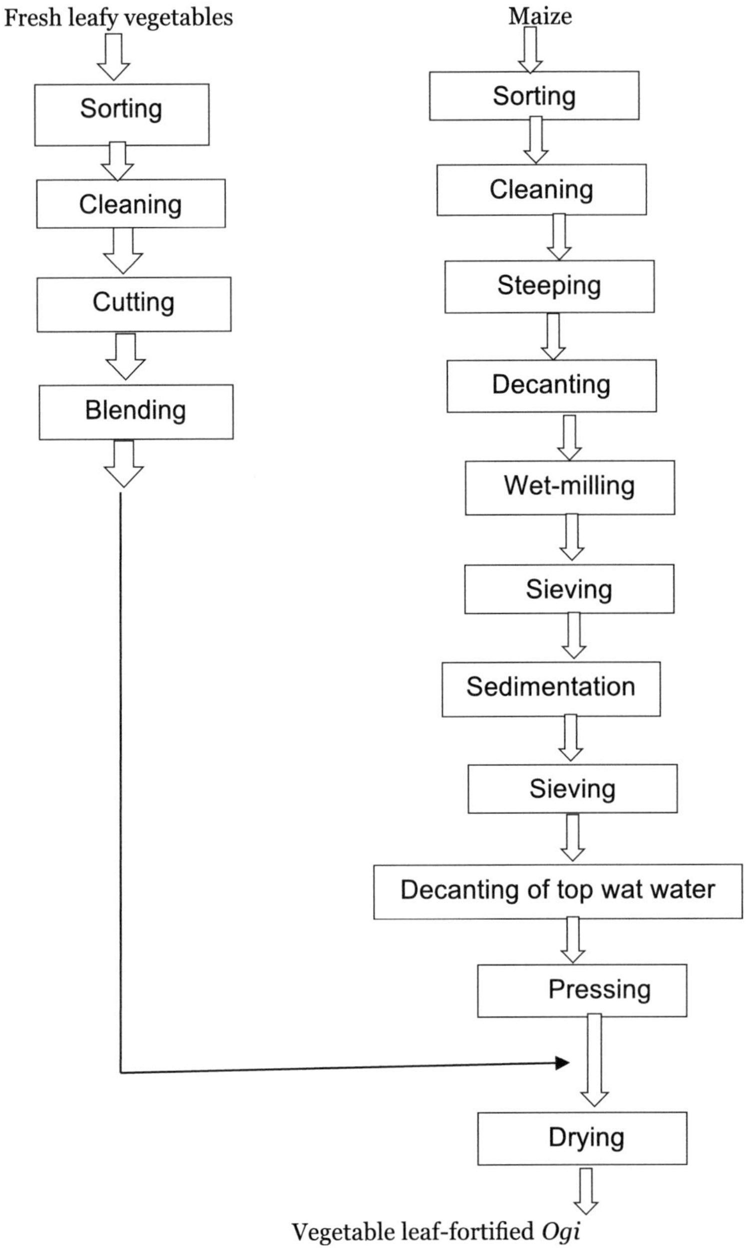

Figure 1.
Process flowchart for the production of vegetable leaf-fortified ogi.

nutritional values. This is because evidence from food consumption surveys in many developing countries has revealed an astronomical rise in the consumption of these floury products, due to changing eating habits, a steadily growing population, and readiness to spend a large proportion of incomes on processed foods [28, 29]. The bread was chosen as a vehicle for vegetable consumption because it is a vital food product that is cherished and consumed all over West Africa. The likeness of bread is not in its nutritional profile but in its sensorial and textural properties, and because it requires no further preparation before consumption. Being a wheat flour product, it contributes greatly to the dietary energy of the population of most West African countries due to its high carbohydrate content [30–32]. Nevertheless, this baked food product is not exceptionally rich in the nutrients that are required for maintaining proper growth and good health, a phenomenon that is attributable to the perceived loss of nutrients during the processing of wheat seeds into flour [33].

As a result of this, research was embarked upon to improve the nutritional base of bread through fortification or enrichment using vegetable-based products [34–37]. Judging from the reported increase in malnutrition and other cardiovascular diseases in less privileged countries of the world, with an ever-increasing interest in the consumption of bread, as a convenience food, there is no other time to improve the nutrient of bread, than now [15]. Improving the nutrient base of bread involves improvement in the protein content and mineral content. It also involves enhancing the radical scavenging ability of the bread, thereby reducing the incidence of the occurrence of some of these degenerative diseases. Therefore, for any material to act as fortificant in bread it must be a good deposit of these nutrients [35, 38, 39]. However, some sensorial properties, such as color, texture, and taste of the fortified bread, were affected by fortification, depending on the plant materials used [40–43]. One of the major challenges being faced in the quest by researchers to fortify bread with these plant materials is the level of inclusion. There is a threshold at which these plant materials can be added to wheat flour during bread making that would preserve both the physical, sensory, and baking quality of the products [44]. Some successes have been recorded with regard to the level at which these materials are included in bread, but this varies with the type of plant materials.

Indigenous leafy vegetables have been shown to possess highly valuable nutritional benefits as a result of the contents of fiber, vitamins, minerals, proteins, and some phytochemicals. These vegetables have been shown to be readily affordable and ready sources of food and nutrients for indigent families in many African countries for a number of years. African population for centuries have been consuming traditional leafy vegetables in large quantities in the preparation of soups and culinary [45, 46]. Based on the information gathered from the literature, there is a dearth of information on the use African leafy vegetables for bread fortification. However, recent MicroVeg works have reported optimization of bread fortification with indigenous African leafy vegetable polyphenolic extracts in addition to the chemical composition and quality characteristics of wheat bread supplemented with *T. occidentalis*, *A. viridis*, and *S. macrocarpon* [15, 27, 29].

The MicroVeg process for the production of vegetable leaf-fortified bread is simple and easily adoptable for homemade bread and industrial purposes. The leafy vegetable to be used for the bread fortification was processed as illustrated in **Figure 2**. The leafy vegetables could be used in two forms: dried powder or as an aqueous slurry, both of which are polyphenol-rich. For the vegetable powder fortified bread, a premix of the wheat flour-vegetable powder was first made. The advantage of this premixing stage is to ensure even distribution of the vegetable powder in the wheat flour. This was done by replacing a known proportion of the wheat flour according to the proportions shown in **Table 1**. After the premixing stage, the required amounts of other standard baking ingredients were added step-wise. All other dry ingredients, such as sugar, salt, and yeast, were added to the mixture. The fat and required quantity of water were added and mixed. Water is a very crucial ingredient in bread baking as it determines the overall quality of the baked product. However, the quantity of water and other

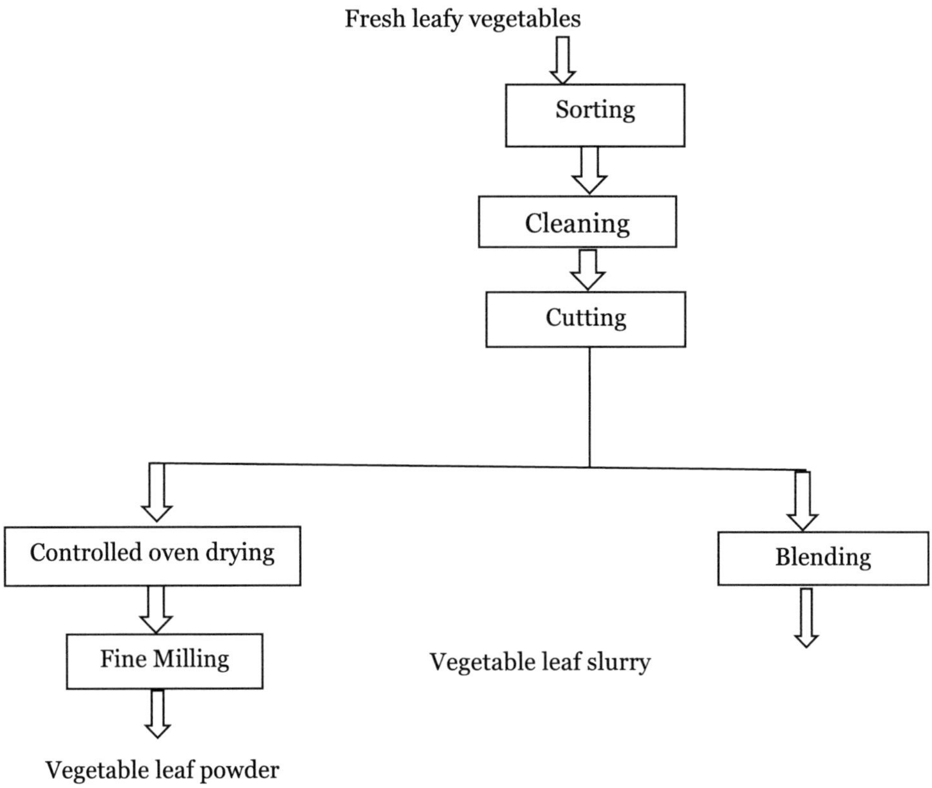

Figure 2.
Process scheme for the production of leafy vegetable dried powder or wet slurry.

standard ingredients needed during bread baking varies, due to differences in the water absorption capacities of different wheat flours. But for vegetable-fortified bread, the standard recipe in **Tables 1** and **2** resulted in good quality vegetable leaf-fortified bread.

For bread fortified with the vegetable leaf slurry, the process differs slightly from that of the dried leaf powder described above. Based on the different levels of vegetable inclusion in the wheat flour (**Table 2**) the water content of the leaf slurry and the balance of water required for blending were taken into consideration. After sorting

Ingredient	Quantity			
	Control	1%	3%	5%
Wheat flour (g)	100	99	97	95
Vegetable powder (g)	—	1	3	5
Sugar (g)	5.00	5.00	5.00	5.00
Yeast (g)	2.00	2.00	2.00	2.00
Margarine (g)	4.00	4.00	4.00	4.00
Salt (g)	1.00	1.00	1.00	1.00
Water (ml)	60.00	60.00	60.00	60.00

Table 1.
Formulation of wheat-vegetable powder composite flour for bread fortification.

Ingredient	Quantity			
	Control	1%	3%	5%
Wheat flour (g)	100.00	99.00	97.00	95.00
Vegetable slurry (g)	—	14.28	42.90	71.43
Sugar (g)	5.00	5.00	5.00	5.00
Yeast (g)	2.00	2.00	2.00	2.00
Margarine (g)	4.00	4.00	4.00	4.00
Salt (g)	1.00	1.00	1.00	1.00
Water (ml)	60.00	48.00	24.00	—

Table 2.
Formulation of wheat-vegetable slurry composite flour for bread fortification.

the vegetable leaves, a slurry was prepared with the understanding that the moisture content of fresh leaves is ~86% (dry matter = 14%). Blending about 100 g of fresh leafy vegetable requires 100 ml of water to obtain a smooth vegetable slurry (200 ml). Of the 200 ml slurry, 186 ml is water, and the solid content is 14 g. With this knowledge, the 1%, 3%, and 5% vegetable leaf slurry to be added during the preparation of bread translated to 14.3 g, 42.9 g, and 71.4 g, respectively. On the basis of 86% (average moisture content of vegetables), ~12.3, 36.9 and 61.4 ml water will be present in 14.3 g (1%), 42.9 g (3%), and 71.4 g (5%) vegetable slurries, respectively. Since water is required during the mixing stage to form an acceptable dough, the quantity of water in the slurry was deducted from the water required. This is necessary in order to establish a basis for comparing the nutritional values of dried vegetable leaf-fortified bread with slurry-fortified bread. In vegetable bread production, the optimum mixing time and proofing time proposed by Famuwagun et al. [29] resulted in good quality vegetable-fortified bread. The products were then baked according to the conditions highlighted in **Figure 3** and allowed to cool to room temperature before packaging to ensure good quality vegetable-fortified bread as shown in figure.

2.5 Vegetable leaf-fortified pastry products

Consumption of pastry products, such as snacks (cookies, *Chinchin*), sandwiches, and burgers, especially among children is on the rise in most West African countries because they are affordable, ready-to-eat, attractive, and have long shelf life [46]. The importance of these pastry products to humanity goes beyond convenience eating but is sometimes used in various feeding programs for school children and areas where there are natural disasters [47]. The MicroVeg project, therefore, concentrated a major effort to develop the fortification of these pastry products as a means of improving their nutritional qualities to meet the health-promoting needs of women and children.

Various plant products that are rich in micronutrients, vitamins, protein, and fiber have been used by researchers to fortify these pastry products with a view to improving their nutritional qualities. Adegunwa et al. [48] improved the nutritional composition of *Chinchin* through the addition of millet flour. The authors showed that up to 30% wheat flour was replaced with millet flour without a pronounced effect on the physical properties of the snack but with improved nutritional qualities. Shrestha and Noomhorm [49] produced high-protein cookies using composite flours that contained

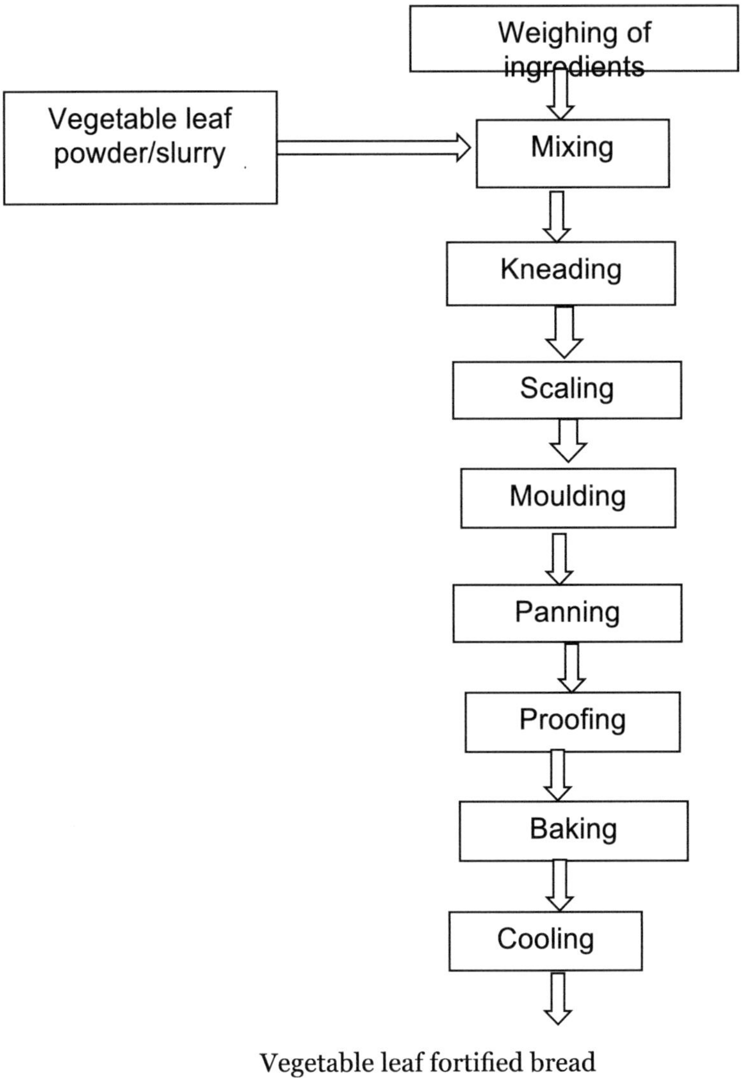

Figure 3.
Process flowchart for the production of vegetable leaf-fortified bread.

blends of wheat-soy flour, defatted peanut, and pea flour, replacing up to 30% of wheat flour without any marked changes in the physical properties of the final products. Cowpea and peanuts flours have been reported to replace about 20% of wheat flour in the production of highly nutritious cookies without any significant effects on the physical characteristics of the products [46, 50, 51]. There are also reports of the production of high-fiber cookies using sesame seeds and pigeon pea flours [52]. Therefore, fortification with indigenous vegetable leaf powder is one of the new approaches aimed at improving the nutritional composition of pastry products.

Cookies and *Chinchin* were fortified with leafy vegetables. The major difference in the production technique of these two commonly consumed pastries lies in the final stage of production. While vegetable cookie doughs were baked, the *Chinchin* dough was fried; therefore, the equipment used and their operating conditions vary. For the two products, it is important to premix the wheat flour and the vegetable powder to be used in fortifying the product. This is to ensure the even distribution of the vegetable powder in the wheat flour matrix. Other ingredients, such as salt, flavoring (nutmeg), margarine, and baking powder (for cookies), can be added after the premixing stage and properly mixed to fine particles without any lumps. Since vegetable-fortified pastries differ from conventional pastries, due to the addition of vegetables, it is important to have an accurate measurement of these ingredients to ensure good quality of the final products. The sweetener (sucrose) was dissolved in the whipped egg (whole egg for *Chinchin,* while only egg yolk is used for cookies) to ensure proper dissolution before adding to the mixture [14]. Beaten eggs or whipped eggs are necessary in order to give the dough a light and airy texture. Liquid milk can be added or in powdered form (must be dissolved with water before use) and then the required amount of water is added to ensure good consistency of the final products. After proper mixing of all the ingredients, the dough was then kneaded. For cookies, the dough was allowed to rest for about 20 min to allow the baking powder to release enough aerating gas before shaping. The cookie dough was then properly shaped into desired sizes and baked for 20 min at 180°C, while the *Chinchin* dough was cut into desired pieces and fried 10 min in hot (~180°C) vegetable oil. The fried *Chinchin* was removed; oil was drained off and the product was allowed to cool down to room temperature. The baked cookie and fried *Chinchin* were separately packaged in air-tight containers.

3. Outcome of the studies

3.1 Sensory evaluation and acceptability of vegetable leaf-fortified bread

Sensory evaluation deals with human sensory perception and their effective responses to food and its components. Sensory evaluation of food products is a function of five major sense organs; the sense of sight, smell, taste, touch, and ear to answer questions related to the preference of the food products under evaluation [53, 54]. In the sensory test, panelists, who are familiar with the products under evaluation are given a range of samples and asked to rate the samples by choosing a point on a scale ranging from "*dislike extremely*" to "*like extremely*" for different attributes of the products. The general sensory attribute of food products includes taste, flavor, aroma, texture, and appearance. Taste is one of the main attributes for evaluating the sensory quality of vegetable fortified bread since the addition of vegetables has the potential on changing the nutritional bread. We evaluated the sensory attributes and acceptability of the vegetable-fortified bread at 41 locations in Nigeria with 250 panelists per location, for a total of 10,250 panelists. As we have reported, taste preference for the bread decreased as the level of inclusion of vegetables increased in the bread [16]. The results suggest that the vegetable taste could be masked by the sweet taste of sugar but this effect diminishes as the vegetable fortification level increased from 1% to 3%.

Color of food products is one of the important attributes for acceptability of any baked product, especially vegetable-fortified bread. Leafy vegetables are green and this color becomes imparted on the dough as the fortification level is increased in the bread formulations. The intensity of the green color on the vegetable-fortified bread depends on the type of vegetable used (**Figure 4**). In the study by Odunlade et al. [16], vegetable bread-fortified with fluted pumpkin leaf (*Telfaria occidentalis*) was preferred up to the 3% inclusion level, while bread-fortified with 1% *Amaranthus* vegetable was mostly preferred compared with the higher level of inclusions (**Table 3**). This is an indication that the likeness of vegetable-fortified bread also depends on the type of vegetable used.

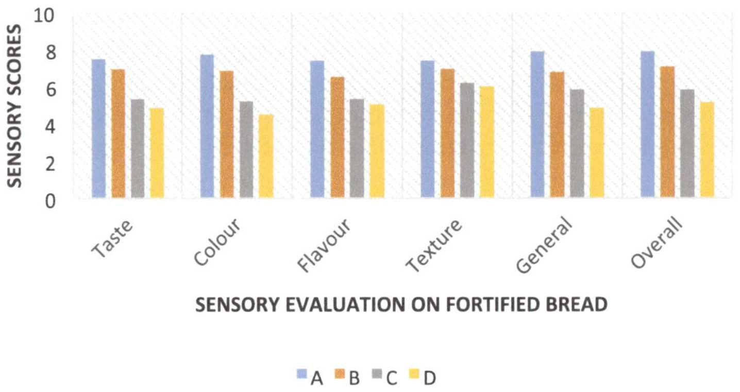

A: control; **B**: 1% leaf slurry; **C**: 3% leaf slurry; **D**: 5% leaf slurry

Figure 4.
Sensory evaluation on bread with fluted pumpkin leaf slurry.

	Taste	Color	Flavor	Texture	Appearance	Overall acceptability
99% WF: 1% TOL	7.5 ± 0.45c	8.0 ± 0.12b	6.0 ± 0.14c	7.3 ± 0.04b	7.2 ± 0.16b	7.9 ± 0.05b
98% WF: 2% TOL	6.4 ± 0.05f	7.7 ± 0.14c	5.5 ± 0.07e	6.9 ± 0.04d	6.7 ± 0.05d	7.2 ± 0.06c
97% WF: 3% TOL	6.2 ± 0.11 g	5.0 ± 0.10f	5.0 ± 0.08 g	6.1 ± 0.05 g	5.0 ± 0.08 g	4.0 ± 0.04j
99% WF: 1% AVL	7.7 ± 0.08b	6.2 ± 0.19d	5.9 ± 0.02d	7.0 ± 0.13c	7.0 ± 0.03c	6.5 ± 0.06d
98% WF: 2% AVL	6.9 ± 0.09e	5.5 ± 0.12e	5.4 ± 0.04f	6.3 ± 0.10f	6.2 ± 0.06e	5.9 ± 0.06 g
97% WF: 3% AVL	5.9 ± 0.31 h	4.9 ± 0.45 g	5.2 ± 0.09 g	6.0 ± 0.04 h	4.8 ± 0.05i	4.4 ± 0.05i
99% WF: 1% SML	7.2 ± 0.21d	6.2 ± 0.37d	6.2 ± 0.03b	7.0 ± 0.04c	6.6 ± 0.03d	6.1 ± 0.06e
98% WF: 2% SML	6.9 ± 0.12e	5.4 ± 0.12e	5.4 ± 0.05e	6.7 ± 0.10e	5.3 ± 0.05f	6.0 ± 0.05f
97% WF: 3% SML	5.1 ± 0.22i	4.8 ± 0.25 h	4.9 ± 0.14 g	6.1 ± 0.04 h	4.9 ± 0.04 h	5.0 ± 0.07 h
100% WF	8.2 ± 0.12a	7.00 ± 0.21a	8.5 ± 0.31a	8.5 ± 0.07a	7.5 ± 0.07a	8.7 ± 0.04a

Mean values along the same column with different superscripts are significantly different ($P < 0.05$)*; WF: heat flour; TOL:* T. occidentalis *leaf powder; AVL:* A. viridis *leaf powder; SML:* S. macrocarpon *leaf powder.*

Table 3.
Sensory properties of wheat bread fortified with green leafy vegetable powders.

Studies have shown that consumer acceptance of a product can be enhanced through additional information on potential health benefits or when the actual health benefits have been proven [55, 56]. For the green bread, consumer acceptance improved when potential health benefits, such as antioxidant effects, cardiovascular health, and blood glucose control, were included in the information package.

3.2 Sensory evaluation and acceptability of vegetable leaf-fortified weaning food

The evaluated sensorial quality attributes were taste, color, flavor appearance, and acceptability. We engaged 100 participants comprising nursing parents, elderly, and university undergraduates who currently utilize maize porridge as weekly meals during the sensory evaluation of the cooked vegetable leaf-fortified weaning food. For all the evaluated sensory parameters, the samples without the addition of leaf extracts were the most acceptable. This was also observed by Aminigo and Akingbala [23] for non-supplemented maize *ogi* samples. It was however observed that the preference for the fortified *ogi* samples ranged from *liked moderately* to *liked slightly*. The maize porridge samples with the inclusion of 10 ml vegetable leaf extract (10 ml of leaf extract to 100 g of *ogi* slurry) were the most preferred among the fortified samples.

With increased inclusion of vegetable leaf extract to maize porridge from 10 to 100 ml, the preferences of the samples for taste, color, flavor, texture, appearance, and general acceptance were observed to significantly ($P < 0.05$) reduce, though within the acceptance range of hedonic scales for sensory analysis. This same trend was reported by Abioye and Aka [18] for moringa fortified maize porridge and Aminigo and Akingbala [18] for okra fortified maize porridge. There were however no significant differences on all the evaluated sensory parameters between the inclusion of 25, 50, and 75 ml of vegetable leaf extract in maize porridge samples. Based on all the sensory parameters, it was further observed that the samples with the incorporation of 100 ml vegetable leaf extract to maize porridge were *neither liked* nor *disliked* by the panelists. It is noteworthy that the addition of vegetable leaf extract to the popular weaning food did not result in rejection of the products by the sensory panelists. However, the better acceptability of the non-fortified product is suspected to be due to the age-long preference and adaption to it, while the change in the color and taste of the fortified products reduced the degree of acceptability. However, a rigorous system of advocacy and awareness creation about the food value of the vegetable leaf extract-fortified weaning food is expected to promote better adoption and utilization of the products by the populace.

3.3 Sensory evaluation and acceptability of vegetable leaf-fortified pastry products

In sensory evaluations of vegetable-fortified pastry, the quality attributes of interest are color, flavor, taste, texture, appearance, and acceptability. In the study conducted by Fasogbon et al. [14] on the quality attributes of *Chinchin* and cookie enriched with *Telfaria occidentalis*, *Amaranthus viridis,* and *Solanum macrocarpon* leaf powders, wheat flours were substituted up to 10% with the vegetable powders. It was observed from the studies that an increase in the level of inclusion of the leaf powders decreased the level of acceptability of the products. Cookies and *Chinchin* fortified with 2% and 5% leaf powder were not significantly different in taste when compared with non-fortified equivalents. In a similar study conducted by Akindele [57], wheat flour was replaced with 1%, 3%, and 5% *Basella alba* and *Telfaria occidentalis* leaf powders in the production of high-quality and acceptable *Chinchin* and cookies. In the study, it was observed that incorporation of the leaf powders did not affect the taste of the snacks when compared with cookies and *Chinchin* produced with 100% wheat flour. Replacement of wheat flour with vegetable leaf powders up to 5% did not affect flavor and texture of the cookies and *Chinchin* [57]. Results from the MicroVeg project are superior to those reported for *Chinchin* and cookies fortified with 5% soybean, which had very low consumer acceptance in terms of flavor and taste [49]. The poor consumer acceptance of the soybean-fortified pastries was attributed to the imparted beany flavor. The green color of vegetable leaf-fortified pastry increased significantly as the inclusion level increased. However, despite the green color, vegetable leaf-fortified pastries were acceptable to consumers to the 5% level of incorporation [14, 57].

3.4 Training of local population on vegetable fortification of foods

The MicroVeg value addition team with the support of the extension personnel trained a total of 5466 persons (68% women) across 25 locations on the promising value addition innovations between September 2016 and July 2017. The training focused on various aspects of vegetable leaf drying, especially the use of a locally designed charcoal-fired system and the production of vegetable-enriched products based on practical demonstrations and video instructions. During the practical sessions, women had hands-on experience producing some of the vegetable-fortified pastry products. The team also trained 100 bakery owners in southwest Nigeria on the practical production of vegetable fortified bread. A combination of video and practical demonstrations of the entire process of vegetable fortification of baked products was used during the training exercise. Participants at the different training sessions also participated in sensory evaluation of the bread and pastry products. In order to create awareness on the nutritional values of vegetables and vegetable-fortified products, the MicroVeg project sponsored daily enlightenment programs on local radio stations. The radio programs reached an estimated 5 million listeners every day in southwest Nigeria. For enhanced technology transfer to the local population (especially women in rural areas), the recipes for green bread and vegetable leaf-fortified snacks have been translated into the local language (Yoruba), the main spoken language in Southwest Nigeria (**Table 4**).

4. Conclusions

Results from the value-addition component of the MicroVeg project have led to the development of novel vegetable leaf-fortified bread and pastry formulation with higher contents of nutrients and bioactive polyphenols. The work also produced a vegetable leaf-enriched traditional maize-based breakfast hot cereal with superior nutritional value than regular food. These vegetable-enriched products constitute an important component of the diets of women and children in Nigeria. Therefore, the availability of these vegetable-enriched food products could enhance maternal and child nutrition through improved nutrient supply in addition to a healthier status due to the free radical scavenging ability of the incorporated polyphenolic compounds. Extension activities led to the training of rural thousands of women and local bakers on the technology for the production of fortified food products. Additional activities

Sample	Color	Aroma	Taste	Mouth feel	Acceptability
A	8.10 ± 0.74^{a}	7.10 ± 0.57^{a}	7.40 ± 1.07^{a}	6.60 ± 1.07^{a}	7.90 ± 0.74^{a}
B	7.20 ± 1.40^{b}	7.00 ± 1.05^{a}	7.40 ± 0.70^{a}	6.70 ± 0.48^{a}	7.30 ± 1.49^{a}
C	6.50 ± 0.53^{b}	7.10 ± 0.57^{a}	7.00 ± 0.69^{ab}	7.30 ± 0.82^{a}	6.90 ± 0.74^{a}
D	5.60 ± 1.07^{c}	6.80 ± 1.22^{a}	6.80 ± 0.79^{ab}	6.70 ± 0.67^{a}	5.70 ± 1.34^{b}
E	4.30 ± 0.67^{d}	6.70 ± 1.49^{a}	6.20 ± 1.32^{b}	6.80 ± 0.92^{a}	5.00 ± 1.33^{b}

The mean values along the same row with different superscripts are significantly different (P 0.05) using Duncan multiple range test. A—100% wheat Chinchin; *B—*Chinchin *enriched with 1% vegetable polyphenol; C—*Chinchin *enriched with 2% vegetable polyphenol; D—*Chinchin *enriched with 3% vegetable polyphenol; E—*Chinchin *enriched with 5% vegetable polyphenol.*

Table 4.
Mean sensory score of chinchin *enriched with polyphenol from* Amaranthus viridis *L.*

are underway to obtain regulatory approval for these vegetable-fortified food products with the aim of eventual commercialization of the respective technologies.

Source of funding

Canadian International Development Research Center (IDRC) and Global Affairs Canada (GAC) through the Canadian International Food Security Research Fund (CIFSRF) program (project 107983).

Acknowledgements

We acknowledge the financial support of IDRC and GAC through Grant No 107983 (Micro-Veg Project).

Conflict of interest

The authors declare no conflict of interest.

Author details

Albert A. Famuwagun[1,2*], Odunayo C. Adebooye[3], Tunji V. Odunlade[2], Kehinde A. Taiwo[2], Durodoluwa J. Oyedele[4] and Rotimi E. Aluko[5]

1 Department of Food Science and Technology, Joseph Ayo Babalola University, Ikeji-Arakeji, Osun State, Nigeria

2 Department of Food Science and Technology, Obafemi Awolowo University, Ile-Ife, Osun State, Nigeria

3 Department of Agronomy, Osun State University, Oshogbo, Osun State, Nigeria

4 Department of Soil and Land Resources Management, Faculty of Agriculture, Obafemi Awolowo University, Ile-Ife, Osun State, Nigeria

5 Department of Food and Human Nutritional Sciences, University of Manitoba, Winnipeg, MB, Canada

*Address all correspondence to: akinsolaalbert@gmail.com

References

[1] WHO/FAO. Diet, Nutrition and the Prevention of Chronic Diseases. Report of a Joint FAO/WHO Expert Consultation, WHO Technical Report Series 916. Geneva: World Health Organization; 2002

[2] Lim SS, Vos T, Flaxman AD, Danaei G, Shibuya K, Adair-Rohani H, et al. A comparative risk assessment of burden of disease and injury attributable to 67 risk factors and risk factor clusters in 21 regions, 1990–2010: A systematic analysis for the global burden of disease study. Lancet. 2012;**380**:2224-2260. DOI: 10.1016/S0140-6736(12)61766-8

[3] Adebooye OC. "African indigenous vegetable resources in a changing world" Promoting green economy. In: Ayuk ET, Oku EE, editors. Implications for Natural Resources Development. The United Nations University; 2014

[4] Aune DE, Boffetta LT, Fadnes N, Keum T, Norat DC, Greenwood E, et al. Fruit and vegetable intake and the risk of cardiovascular disease, total cancer and all-cause mortality—A systematic review and dose-response meta-analysis of prospective studies. International Journal of Epidemiology. 2017;**46**: 1029-1056. DOI: 10.1093/ije/dyw319

[5] WHO. Promoting Fruit and Vegetable Consumption around the World. Geneva: World Health Organization; 2015 https://www.who.int/dietphysicalactivity/fruit/en/. [Accessed: November 29, 2017]

[6] Nestel P. Food fortification in low-income countries: A new approach to an old standby. In: Food Fortification in Developing Countries U.S. Washington, DC: Agency for International Development; 1993

[7] West CE, Eilander A, van Lieshout M. Consequences of revised estimates of carotenoid bioefficacy for dietary control of vitamin a deficiency in developing countries. Journal of Nutrition. 2002, 2002;(132):2920S-2926S

[8] FAO/WHO. Development of Guidelines on the Fortification Requirements of Lower Fat Products. Rome: FAO; 1995; CX/NFSDU 95/9

[9] FAO. Preventing micronutrient deficiencies: Food abundance and diversity are fundamental. Food, Nutrition and Agriculture. 1993;**1993**(7): 8-17

[10] Horton S. The economics of food fortification. The Journal of Nutrition. 2006;**36**:1068-1071

[11] Adebooye OC, Alashi MA, Aluko RE. Economic potentials of polyphenols. Journal of Food Biochemistry. 2017; **2017**(1):21-29

[12] Famuwagun AA, Taiwo KA, Gbadamosi SO, Oyedele DJ, Aluko RE, Adebooye OC. Extraction optimization and antioxidant properties of African eggplant (*Solanum macrocarpon*) leaf polyphenols. Journal of Food Quality. 2017; Article ID 2159183. DOI: 10.1155/2017/2159183

[13] Alashi AM, Taiwo KA, Oyedele DJ, Adebooye OC, Aluko RE. Blood pressure lowering ability of vegetable-fortified bread. In Afri-veg Forum, 2017 Forum on African Vegetables, Cotonou, November 20–22, 2017. Acta Horticulture. 2017

[14] Fasogbon BM, Taiwo KA, Oyedele DJ. Nutritional assessment and consumer acceptability of snacks

(chinchin and cookies) enriched with underutilized indigenous vegetables. International Journal of Food and Nutritional Sciences. 2017;**6**:97-107

[15] Odunlade TV, Famuwagun AA, Taiwo KA, Gbadamosi SO, Oyedele DJ, Adebooye OC. Chemical composition and quality characteristics of wheat bread supplemented with leafy vegetable powders. Journal of Food Quality. 2017; Article ID 9536716. DOI: 10.1155/2017/9536716

[16] UNICEF. (2017). Malnutrition rates in children under 5 years. 2017. https://www.unicef.org/nigeria/factsheets_NUTRITION_low.pdf . [Accessed November 26, 2017].

[17] Odunlade TV, Taiwo KA, Adeniran HA. Functional and antioxidative properties of sorghum Ogi flour enriched with cocoa. Annals Food Science and Technology. 2016;***17***: 497-506

[18] Abioye VF, Aka MO. Proximate composition and sensory properties of moringa fortified maize-Ogi. Journal of Nutrition and Food Sciences. 2015;**12**: 234–242

[19] Oladeji OA, Taiwo KA, Ishola MM, Oladeji BS. Nutritional and quality characteristics of white maize *Ogi* flour enriched with *Moringa oleifera* seed. Biotechnology Journal International. 2017;**18**:1-11

[20] Ajanaku KO, Ogunniran KO, Ajani OO, James OO, Nwinyi OC. Improvement of nutritive value of sorghum Ogi fortified with pawpaw (*Carica papaya* L.) fruit vegetable cereal science. Biotechnology. 2010;**4**:98-101

[21] Sanni AI, Asiedu M, Ayernor GS. Influence of processing conditions on the nutritive value of *Ogi-baba*, a Nigerian fermented sorghum gruel. Plant Foods for Human Nutrition. 2001;**56**: 217-223

[22] Osundahunsi OF, Aworh OC. Nutritional evaluation, with emphasis on protein quality, of maize-based complementary foods enriched with soya bean and cowpea tempe. International Journal of Food Science and Technology. 2003;***38***:809-813

[23] Aminigo ER, Akingbala JO. Nutritive composition of Ogi fortified with okra seed meal. Journal of Applied Sciences and Environmental Management. 2004; **8**:23-28

[24] Aminigo ER, Ossai GEA. Production and evaluation of high-protein roasted maize meal. Journal of Applied Sciences and Environmental Management. 1999; **3**:17-22

[25] Ijarotimi SO, Bakere SS. Evaluation of proximate, mineral and antinutritional factor homemade processed complementary diet from locally available food materials (*Sorghum bicolor* and *Sphenostylis stenocarpa*). Journal of Food Technology. 2006;**4**: 339-344

[26] Oladeji BS, Akanbi CT, Ibironke SI. Physicochemical and nutritional evaluation of co-processed fermented yellow maize Ogi (an infant diet) and carrot blends. Annals Food Science and Technology. 2014;**15**:82-91

[27] Odunlade TV. Famuwagun AA, Taiwo KA, Gbadamosi SO, Oyedele DJ, Adebooye OC. Development of maize ogi fortified with fluted pumpkin leaf polyphenolic rich extract. Microveg/Afriveg Forum, 2017. African Vegetables: From Pot to Plate, November 19–22, Sun Beach Hotel, Cotonou, Benin Republic 91.

[28] Anyika JU, Uwaegbute AC. Frequency of consumption and nutrient content of some snacks eaten by an adolescent secondary and university student in Abia state. Nigerian Journal of Nutritional Sciences. 2005;***26***:10-15

[29] Famuwagun AA, Taiwo KA, Gbadamosi SO, Oyedele DJ. Optimization of production of bread enriched with leafy vegetable powder. Journal of Food Processing and Technology. 2016;*7*:7

[30] Onoja US, Dibua UME, Eze JI, Odo GE. Physico-chemical properties, energy, mineral, vitamin and sensory evaluation of wheat-based bread supplemented with legume, root, tuber and plantain flour. Global Journal of Pure and Applied Science. 2011;**17**: 319-327

[31] Akubor PI. Influence of germination and fermentation on the qualities of bread prepared from cowpea and wheat flour blends. Nigeria Journal of Nutritional Science. 2008;**29**:156-166

[32] Mastromatteo M, Danza A, Lacce L, Spinelli S, Lampignano V, Laverse J, et al. Effect of durum wheat varieties on bread quality. International Journal of Food Science and Technology. 2013; **2013**(49):72-81

[33] WHO/FAO. The joint World Health Organization/food and agricultural organization expert consultation on diet, nutrition and the prevention of chronic diseases: Process, product and policy implications. Public Health Nutrition. 2004;*7*(*1A*):245-250

[34] Peng X, Ma J, Cheng KW, Jiang Y, Chen F, Wang M. The effects of grape seed extract fortification on the antioxidant activity and quality attributes of bread. Food Chemistry. 2010;**119**:49-53

[35] Ranawana DV, Raikos V, Campbell F, Bestwick C, Nicol P, Milne L, et al. Breads fortified with freeze-dried vegetables: Quality and nutritional attributes. Part 1: Breads containing oil as an ingredient. Food. 2016;**5**:19

[36] Świeca M, Gawlik-Dziki U, Dziki D, Baraniak B, Czyż J. The influence of protein–flavonoid interactions on protein digestibility in vitro and the antioxidant quality of breads enriched with onion skin. Food Chemistry. 2013; **141**:451-458

[37] Świeca M, Sęczyk L, Gawlik-Dziki U, Dziki D. Bread enriched with quinoa leaves—The influence of protein–phenolics interactions on the nutritional and antioxidant quality. Food Chemistry. 2014;***162***:54-62

[38] United States Department of Agriculture and States Department of Agriculture and United States Department of Health. Dietary Guidelines for Americans. 7th ed. United Washington, DC, USA; 2010

[39] Ogunsina B, Radha C, Indrani D. Quality characteristics of bread and cookies enriched with debittered *Moringa oleifera* seed flour. International Journal of Food Science and Nutrition. 2010;***62***:185-194

[40] Ameh MOI, Gernah D, Igbabul BD. Physico-chemical and sensory evaluation of wheat bread supplemented with stabilized undefatted rice bran. Journal of Food and Nutrition Sciences. 2013;**4**: 43-48

[41] Igbabul B, Num G, Amove J. Quality evaluation of composite bread produced from wheat, maize and orange fleshed sweet potato flours. American Journal of Food Science and Technology. 2014;*2*: 109-115

[42] Bolarinwa IF, Aruna FE, Raji AO. Nutritive value and acceptability of bread fortified with moringa seed powder. Journal of the Saudi Society of Agricultural Sciences. 2017. DOI: 10.1016/j.jssas.2017.05.002

[43] Abdelghafor RF, Mustafa AI, Ibrahim AMH, Krishnan PG. Quality of bread from composite flour of sorghum and hard white winter wheat. Advanced Journal of Food Science and Technology. 2011;**3**:9-15

[44] Gellynck X, Kühne B, Van Bockstaele F, Van de Walle D, Dewettinck K. Consumer perception of bread quality. Appetite. 2008;**53**:16-22

[45] George P. Encyclopedia of Foods. Vol. 1. Washington: Humane Press; 2003. p. 526

[46] Noorfarahzilah M, Lee JS, Sharifudin MS, Mohd Fadzelly AB, Hasmadi M. Applications of composite flour in development of food products. International Food Research Journal. 2014;***21***:2061-2074

[47] Hooda S, Jood S. Organoleptic and nutritional evaluation of wheat biscuits supplemented with untreated and treated fenugreek flour. Food Chemistry. 2005;**90**:427-435

[48] Adegunwa MO, Ganiyu AA, Bakare HA, Adebowale AA. Quality evaluation of composite millet-wheat Chinchin. Agriculture and Biology Journal of North America. 2014. DOI: 10.5251/abjna.2014.5.1.33.39

[49] Shrestha AK, Noomhorm A. Comparison of physicochemical properties of biscuits supplemented with soy and kinema flours. International Journal of Food Science and Technology. 2002;**37**:361-368

[50] McWatters KH, Ouedraogo JB, Resurrection AVA, Hung YC, Philips RD. Physical and sensory characteristics of sugar cookies containing mixtures of wheat, fonio (*Digitaria exilis*) and cowpea (*Vigna unguiculata*) flours. International Journal of Food Science and Technology. 2003; **38**:403-410

[51] Akubor PI. Protein contents, physical and s lensory properties of Nigerian snack foods (cake, chin-chin and puff-puff) prepared from cowpea-wheat flour blends. International Journal of Food Science and Technology. 2004;**9**: 419-424

[52] Piteira MF, Maia JM, Raymundo A, Sousa I. Extensional flow behaviour of natural fibre-filled dough and its relationship with structure and properties. Journal of Non-Newtonian Fluid Mechanics. 2006;**137**:72-80

[53] Carpenter RP, Lyon DH, Hasdell TA. Guidelines for Sensory Analysis in Food Product Development and Quality Control. Springer Open; 2000

[54] Tuorila H, Cardello AV. Consumer responses to an off-flavor in juice in the presence of specific health claims. Food Quality and Preference. 2002;**13**:561-569

[55] Frewer L, Scholderer J, Lambert N. Consumer acceptance of functional foods: Issues for the future. British Food Journal. 2003;**105**:714-731

[56] Urala N, Lähteenmäki L. Consumers 'changing attitudes towards functional foods. Food Quality and Preference. 2007;**18**:1-12

[57] Akindele O. Studies on production of *chinchin* enriched with vegetables. [M. Sc. Thesis]. Ile-Ife, Osun State, Nigeria: Obafemi Awolowo University; 2016

Chapter 3

Role of Edible Insects as Food Source to Combat Food Security Challenges – Innovative and Traditional Approaches

Mahnoor Pervez and Farkhanda Manzoor

Abstract

World Food Security organization committee stated that food security exists when safe, healthy, and nutritious food is present to all human beings, to meet their dietary needs. There are four pillars of food security such as availability, access, utilization, and stability, while the nutritional dimension is considered integral to the concept of food security. Protein is an essential dietary component. It is required for the proper growth of individuals. Meat is an excellent source of protein. However, because of the increasing demand for meat and declining agricultural land. There is an urgent need to discover alternative ways to full fill public protein demand. Edible insects are the best alternative source of protein. They can be harvested in large amounts. Insect meals can replace scarce fishmeal as feed ingredients. Insects are not only a source of protein but also contained various nutrients and minerals such as iron, zinc, and selenium. People showed awareness of the importance of insects as an essential dietary element. Edible insect harvesting is a new emerging agriculture sector and needs the special attention of researchers and the government. Thus, insects require the use of commercial processing methods that will render the protein suitable for food/feed formulation, while maintaining the safety, nutritional and sensory quality of the final product.

Keywords: food security, food crisis, starvation, edible insects, cheap protein

1. Introduction

The term "food security" refers to the access to an adequate amount of food for meeting dietary energy needs that implies for many as self-sufficiency as producing required food domestically [1]. It is a basic right of a human being to have access to adequate production, availability, and food consumption to ensure adequate nutrients and food for all the people.

Food security ensures that all human beings have physical and economic accessibility to healthy and nutritious food to meet their daily dietary requirements and maintain an active healthy lifestyle. A country promises food security when it can ensure food availability balance between demand and supply to the nation, no matter what the price would be.

The five fundamental components of food security are availability, access, stability, nutritional status, and preference for food. All of them are prominently influenced by various factors such as political, physical, and economic within a community. Food availability and accessibility are two main aspects to ensure food security. Food security ensures an adequate amount of nutrients for every individual in the country. In many countries an adequate amount of food is available, but its availability is a real serious issue in some areas because of long droughts, political issues, mismanagement of resources, and unfertile land. These issues are most commonly observed in Asian countries.

The term accessibility refers to the capacity to produce and acquire an appropriate amount of healthy food for all individuals [2]. The availability of adequate food at the country/local level does not guarantee that all people have food accessibility and are food secure. Because various aspects such as low income, lack of roads, and infrastructure could restrict the accessibility of basic foodstuff. Hence, both availability and accessibility are essential and inter-linked parts of food security [1].

Stability is another important aspect that refers to a consistent supply of nutritious food at the national level as well as stability in access to food at the household and individual levels. Stability is directly affected by agricultural practices. Stability is ensured by storing food and enhancing shelf life. However, in developing countries economic crises are critical and only a small amount of food is stored for the whole year. Therefore, besides production, food stability ensures by better management, the formation of new laws and rules, market integration, and trade [3].

So, the above discussion concluded that food security ensures by the conformity of various dimensions such as availability, accessibility, stability, nutritional status, and preferences. It highlights the detail that reaching food security is a tough, multifaceted, and challenging phenomenon.

2. Challenges

Challenges in reaching food security points are a hot topic globally, as the world population is expected to touch more than 9 billion by 2050. So, it is an urgent urge to address and recognized the issues associated with food security [4]. It is suggested that to ensure the food security to meet the food demand of the future population, food production will need to see an upward trend by at least 70% in agricultural production. It is studied that globalization and current global environmental changes have a great impact on food production and distribution. Challenges can be coped by addressing the following issues such as,

- Reviving agricultural productivity growth to feed Asia's growing population.
- Ensure small-scale farmers with limited resources should be equipped with modern food technology practices.
- Addressing poor countries' malnutrition challenges in a sustainable and efficient manner.

Furthermore, the recent vulnerable food crisis raises the question of whether existing food policies in this era are sufficient or not. Scientists and researchers argue that finding a reasonably efficient and politically acceptable set of policies regarding food crisis challenges showed be done. All these policies also decrease prices that encourage farmers to respond with supply responses 3w.

2.1 Climate change

Climate changes are greater threat to agriculture and smooth food supply. The rising intensity of rainfall in some areas, hot temperature with drought conditions in other areas, and climate-related disaster shows that alteration in climate conditions have quantifiable effects on agriculture in a wide range of economies, crops, and farming systems [5]. A wide range of edible food crops such as wheat, rice, and corn are grown only within a narrow temperature range, which varies according to developmental stage, and slight fluctuation in temperature extremely affects the quantity and quality of crops. Scientific evidence revealed that as average temperatures rise then short periods of extreme heat will become more frequent and have more severe impacts on yields [6].

Extreme heat will pose a more severe risk to crop end yield than the impacts of variations in rainfall associated. Similarly, the effect of extreme temperature is intensified under drought conditions.

2.1.1 Temperature variations

In order to study the effect of rising temperature on the yield of the crop, a graph has been plotted among both by obtaining the data of corn crops from US countries from 1950 to 2005, which shows nonlinear relation among them as shown in **Figure 1**.

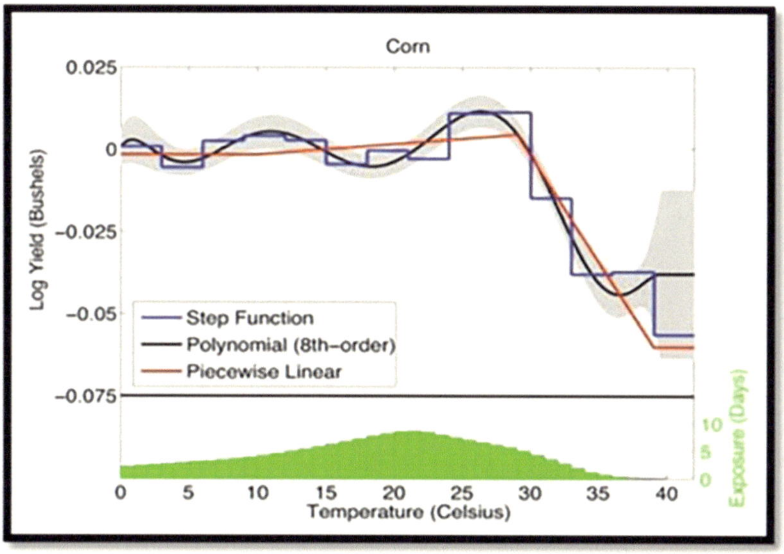

Figure 1.
Effect of rising temperature on crop yield.

2.1.2 Ozone level

Increasing levels of ozone (O3) are another significant source of crop yield loss and an existing threat to Asian food security. In East Asia regions there is a significant increase in ozone level that has been recognized as a drastic factor on major crops and trees. Numerous studies document how global and regional yields of staple crops are being reduced by current O3 levels. Due to the increasing level of O3 in the US, the annual cost of edible crop production lost $4 billion to $2 billion [7]. Similarly, the total cost It has been estimated that the global loss of wheat, rice, corn, and soybean amounted to $14 billion–$26 billion in the year 2000, which is on a par with, and possibly higher than, crop losses presently caused by other aspects of climate change.

Tamburino et al. have discussed the approaches in the scientific literature on the vigorously debated problem of how to feed the world. There seems to be an increasing focus on food production by means of new innovative technologies [8].

2.2 Edible insects as protein source

Acheta domesticus, A. mellifera (honeybee), *Bombyx mori* (farmed silkworm), *Imbrasia belina* (mopane caterpillar), *Tenebrio molitor* (yellow meal worm), and *Rhynchoporus phoenicis* (African palm weevil) are the six most prevalent commercial edible insect species at the moment. The edible ratio of different orders is shown in **Figure 2**. Only the larvae of several species (*R. phoenicis*, *T. molitor*, and *I. belina*) are frequently ingested, as their rich lipids can produce a fascinating flavor. Because of its vigor, *T. molitor* production has been mechanized [9]. Only the adults of orthopterans, such as crickets, are commonly consumed. Swarming makes them extremely effortlessly harvest. Although, because they are common pests in most locations and are frequently exposed to pesticides throughout their lives, there is concern that may compromise food safety, which is the presence of latent pesticides.

Ingesting insects in a responsible and healthy manner has become a popular way to alleviate poverty. Insects that are edible play a vital part in a variety of food systems. In fact, in many underdeveloped countries, they have provided a critical supply of

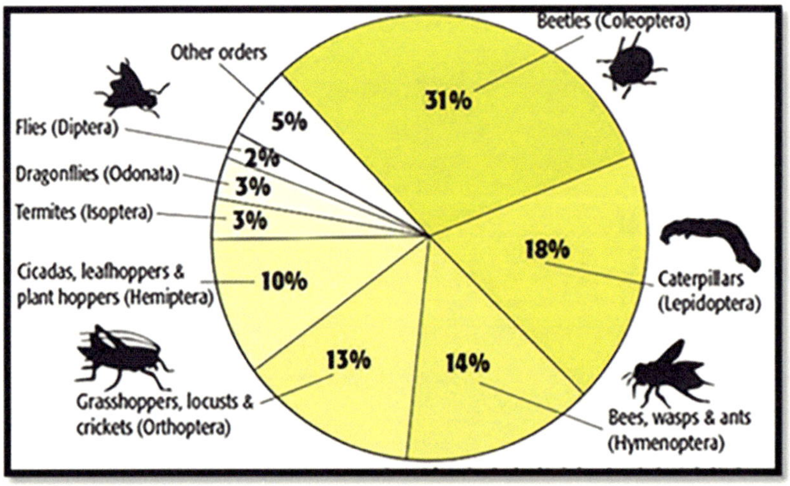

Figure 2.
Edible insect ratio of different orders.

essential nutrients. Insects have gained much attention in food sector because of their protein enrichment, high mineral, and vitamin content, and the unique ratio of polyunsaturated to saturated fatty acid value [10]. In most cases, edible insect proteins meet WHO essential amino acid content standards [9]. Furthermore, the digestive capacity of insect proteins is more than plant-based proteins like peanuts and lentils (53%), but less than animal-based proteins like beef and egg have 100% capacity.

Insect proteins are particularly appealing because they produce less greenhouse gas and ammonia, perhaps making them more environmentally benign [11]. Furthermore, because they are relatively quick, easy, and cheap to farm, they provide a new and accessible commercial potential for underdeveloped countries, making them the overall best alternative for ensuring food security with the world's ever-growing population [12]. Insect proteins offer a lot of good qualities, such as excellent nutritional content, and digestibility, and they are also environmentally beneficial. Furthermore, for including them in a variety of foods, their functional qualities must be properly evaluated at each stage of the development phase. In general, insects have apparent nutritional advantages. Their nutritional profiles are remarkably comparable to that of traditional animal meals [13]. Not only for humans but also for poultry, they offer huge potential as a source of nutrients and active chemicals.

Animal protein is abundant in all phases of the life of insects. It has been proposed that insects are significantly more sustainable when compare with other sources of animal protein, thus alleviating the pressure on the environment and the planet facing the need to full fill the feed the worldwide population, constantly increasing [14]. Bioactive compounds such as antioxidant agents, insulin regulators, and anti-inflammatory peptides are high-value products that can be obtained from insects [15].

The amino acid profiles of edible insects vary, which is indicated in **Table 1**. Although their amino acid contents are usually similar, edible insects typically contain more crude protein than regular meat. They can deliver vital amino acids at an appropriate level as food, with a digestibility of 76 to 96 percent [20]. The World Health Organization has reported the amino acid composition of certain edible insects (WHO 2007). Methionine, cysteine, and tryptophan are all missing or present in extremely little amounts in some insects. If these insects make up the majority of a meal, the diet must be balanced. Aside from these species, insects generally meet the WHO's amino acid recommendations. By ingesting a suitable variety of items, the majority of them can offer enough amounts of essential amino acids. With enough isoleucine, leucine, lysine, phenylalanine, threonine, valine, arginine, histidine, and tyrosine, all groups are qualified. When compared to other insects, Blattodea has the highest concentration of lysine, valine, methionine, arginine, and tyrosine. Coleoptera has a higher concentration of leucine than other protein sources, including animals. Similarly, Hemiptera has a higher concentration of phenylalanine than any other known protein source. Nymphs are usually the most plentiful source of nearly all sorts of amino acids when compared to edible insects at other stages [12]. They possess high arginine content that strengthens the immune system and enhances blood vessel and health function. Beef and pork are more prevalent in arginine as it is in cockroach nymphs i.e., *Blatta lateralis*.

Insects have a lot of lipids in them. The fatty acid content of several edible insects varies greatly. Aside from the acids indicated in **Table 2**, trace amounts of additional acids, such as odd-numbered fatty acids, even-numbered saturated fatty acids, and some unsaturated fatty acids, have been detected in some insects. These acids are thought to be insignificant. On a dry weight basis, the fat content of insects in embryonic stages ranges from 8 to 70%. Different sources of meat, including all groups of

Species	Stage	Crude protein %	Essential amino acids							
			Ile	leu	lys	Met	Phe	Thr	Trp	Val
C. barbatus	Larvae	41.0	5.80	10.0	5.70	2.00	4.70	4.00	0.70	7.00
Tenebrio molitor	Larvae	48.0	4.00	6.90	4.90	1.20	3.20	3.60	1.00	5.90
Hermetia illucens (Soldier fly)	Larva	49.0	4.00	6.60	5.60	1.40	3.80	3.60	1.10	5.60
M. domestica (House fly)	Pupa	62.0	3.50	5.30	5.20	2.60	4.20	3.20		3.40
M. domestica	Larva	54.0	3.20	5.70	6.90	2.20	5.00	3.30	3.20	4.40
A. mellifera (honeybee)	Adult	51.0	5.60	9.60	6.60		0.90	5.50	0.00	6.90
A. mellifera	Larva	42.0	6.00	9.40	7.00		0.80	6.10	0.00	6.20
Acheta domesticus (cricket)	Adult	62.0	2.60	4.50	3.50	0.90	1.40	2.20	0.40	3.70
S. histrio (grasshopper)	Adult	60.0	5.30	8.70	5.70	0.70	4.40	4.00	0.60	5.10
Beef			5.10	8.40	8.40	2.30	4.00	4.00		
Pork			4.90	7.50	7.90	2.50	4.10	5.10		
Chicken	Broiler breast		4.20	6.90	7.80	2.10	2.50	3.70		4.60
Amino acid requirement in human nutrition			30.0	59.0	45.0	16.0		23.0	6.0	39.0

Ile Isoleucine, Leu Leucine, Lys Lysine, Met Methionine, Cys Cysteine, Phe Phenylalanine, Tyr Tyrosine, Thr Threonine, Trp Tryptophan, Val Valine, Arg Arginine, His Histidine.
The data were summarized by the following references: [9, 13, 16–19].

Table 1.
Amino acid content of common edible insects (% in crude protein of dry weight).

insects, have identical fatty acid contents [23]. Lepidopteran and Heteropteran larvae have a higher fat content than other edible insects. When compared to insects at other stages, an excellent reservoir of fatty acid is larvae. Adults are generally slender, with a fat level of less than 20%. Triacylglycerol is the most common type of fat found in insects [16]. More than 80% of all fats are made up of saturated fatty acids (SFAs) and monounsaturated fatty acids (MUFAs). Palmitic acid and stearic acid make up the majority of SFAs in insects at various times. Adults' SFA levels are often greater than MUFA levels, despite the latter being healthier for human diets. In insects, oleic acid, a prevalent monounsaturated fatty acid in the human diet, has the highest concentration of MUFAs. It aids in the reduction of blood pressure in humans and has significant potential in the treatment of inflammatory, immunological, and cardiovascular illnesses [24]. In contrast to pork, beef, and insects at previous stages, adult insects are the greatest site of polyunsaturated fatty acids (PUFAs) with the risk of surpassing SFAs as food.

Insects are also excellent providers of vitamins and minerals. Biochemical substances such as vitamins A, B1–12, C, D, E, and K, which are required for normal

Species	Crude fat %	Saturated fatty acids			Monosaturated Fatty acids		Polyunsaturated fatty acids	
		C14:0	C16:0	C18:0	C16:1	C18:1	C18:2	C18:3
Tenebrio molitor Larvae	38.0	0.39	28.2	0.89	5.98	60.6	2.84	0.18
Hermetia illucens Larva	26.0	99.8	13.2	2.02	4.08	12.8	13.9	0.52
M. domestica Pupa	15.5	3.20	27.6	2.20	20.6	18.3	14.9	2.10
M. domestica Larva	24.3	6.83	26.7	2.31	25.9	21.7	16.4	0.00
A. mellifera Adult	12.3	0.60	14.4	9.30	2.6	45.2	7.80	
Bombyx mori pupa	35.0	0.10	24.2	4.50	1.7	26.0	7.30	36.3
Beef		32.2	0.77	16.7	18.8		36.10	6.16
Pork		41.0	3.43	16.7	43.0	2.93	7.29	1.71
Chicken		33.3	1.33	21.6	46.6	0.27	14.0	0.67
Amino acid requirement in human nutrition		26.9		22.6	30.8		14.95	1.50

SFA saturated fatty acids: C14:0, myristic acid; C16:0, palmitic acid; C18:0, stearic acid. MUFA: C16:1 n7 – palmitoleic acid; C18:1 n9 – oleic acid. PUFA: C18:2 n6 – linoleic acid; C18:3 n3 – α-linolenic acid.
The average data of each order are bold. The data were summarized by the following references: [9, 13, 16–18, 21, 22].

Table 2.
Fat content (%) of common edible insects (% in crude fat of dry weight).

growth and health, could be provided by them [17]. Caterpillars, for example, are particularly high in B1, B2, and B6 [9]. Vitamins A and D are abundant in bee brood (pupae) [25]. *Rhynchophorus ferrugineus*, the red palm weevil, is high in vitamin E. [26]. Edible insects include a wide range of micronutrients, including iron, magnesium, manganese, phosphorus, potassium, selenium, sodium, and zinc [9]. Mineral elements are found in varying amounts in different insects. Most insects have a low calcium content (less than 100 mg/g dry matter), while larvae of house flies and adults of melon bugs have a high calcium content. Immature insects such as pupae and larvae are preferred as food sources because they are abundant in amino acids and fatty acids. These nutrients not only give high nutritious value but also provide a unique splendid flavor [27].

In addition to humans, edible insects also provide the best feed source for aquaculture and livestock. Fowls fed on grains are less nutritional than the fowls that are fed on insects, which provide a good protein source [28]. Moreover, they can also be used as a feed source in areas where the vegetable feed is costly [29]. Soy diet feeding of farmed insects is extremely expensive, yet insect food can provide efficient nourishment at minimum cost. Proteins are more sustainable by recycling biomass during insect production [30]. Yellow mealworms have long been utilized as a source of food [31].

However, while edible insects have been marketed as an environmentally benign protein source in the West, eating insects is still seen as a sign of poverty in underdeveloped countries where entomophagy is a traditional practice. This may be seen in many Asian nations, such as Mexico, where insect ingestion is popular in urban and tourist regions but outlawed in rural ones. Despite the high desire for edible insects, their planting and harvesting are still restricted because farmers fail to recognize their nutritional worth [32]. A successful plan must be considered to boost the inclusion of insects in everyday diets.

3. Edible insect farming

The rising demand for edible insects has generated a problem: how to produce more edible insects in a cost-effective, safe, and long-term manner [33]. Due to a shortage of availability, accessibility concerns arise, limiting prospects for increased trade [34]. This necessitates a technological shift from wild harvesting to indoor agriculture. Traditional and scientific understanding is necessary for the supply of insects through large-scale cultivation for this purpose [35].

Harvesting edible insect species at different phases from wild woods, fields, and waterways can provide huge diversity of edible bug species [36]. These cultivators have the knowledge and experience to determine the right timing, circumstances, and host vegetation to obtain precise culinary species while minimizing environmental impact [37]. Insects are primarily taken from fields for personal consumption, which has a minimum impact on the environment and helps keep insect populations stable over time. Practicing traditional insect harvesting has been done in Mexico prior to the pre-Hispanic period when insects were considered as an essential food source rather than pests, and more than 400 wild insect species were taken from terrestrial environments [38]. Harvesting varies by species and is dependent on the level of growth (eggs, pupae, larvae, or adults), season (rain or dry), and destination (forest, desert, or agricultural fields). For example, pupae of the ant species *Limetopum apicuatum* are gathered during the warm dry season (February to May) [39] and when the rainy season begins, grasshoppers of the species *Sphenarium purpurascen*s (chapulines) are physically picked from fields [34].

Many edible insects are affected by agricultural intensification, owing to mechanization, tree clearing, and pesticide use. Insect populations have declined worldwide as a result of pesticides, with a 67 percent reduction in invertebrate numbers in the previous 40 years [40]. If pesticide controls were removed, a specific grasshopper species' potential production in Mexico would earn about US$350,000 in annual income and provide enough protein to feed nearly 9 million people [41].

Changes in the trophic chain are creating ecological degradation as a result of uncontrolled overharvesting [42]. Increased demand for a few insect species has resulted in more aggressive collecting tactics, with little regard for ensuring sustainable collection during reproductive periods—a practice that threatens to decrease or perhaps abolish native biodiversity [33]. Due to the extreme demands of honey ants and wood grubs, the adoption of energy approaches in agriculture could minimize or even eliminate the usage of pesticides, resulting in a positive environmental impact [32].

Because of the environmental and health dangers, more edible wild insect species are being semidomesticated in a more sustainable, less time-consuming, and consistent manner, mostly in traditional consumer countries where the warm, wet weather is suitable for cultivation [43]. Outdoor farming involves breeding insects in old trees, sawdust, and forest wastes, which can provide a semi-confined habitat as well as a food source. Bamboo caterpillars in Thailand mate in nylon net cages suspended across developing bamboo stalks. Bamboo caterpillars are commonly sold in local markets in plastic boxes [33]. Semidomesticated bees are raised in wooden hives, ancient tree trunks, and other containers in Africa and Asia.

Many edible insect species are partially cultivated, but some are fully cultivated for a variety of reasons, including:

- They have a short life cycle.
- They can be easily reared in small containers,
- Can be consume animal manure rather than grains.

4. Processing technologies

The edible insect industry is growing quickly, and supply of new goods is growing every day [44]. The manufacturing process begins with raw insect post-harvesting and finishes with the creation of foodstuffs and wastes, some of which have been retrieved and recovered as by-products [31]. Different methods are employed in the food edible industries, depending on the species used and the ultimate product desired; however, the majority of them may be classified under relatively modest operations such as blanching, drying, and storage.

4.1 Blanching

Blanching is the process in which food is immersed in boiling water for a short period of time, then removing it and placing them into ice water to stop the thermal process [45]. On the industrial and artisanal scale, it is utilized as a pre-treatment for inactive degradative enzymes and to lower bacterial counts that lead to food spoilage and food poisoning. Wynants and his friend submerged *Alphitobius diaperinus* (beetle) in a bath of water heated to 90 degrees celsius till the temperature reached 88 degrees celsius (5 min). As an outcome, the total microbial count is reduced, and pathogens are missing [46].

Blanching diminishes quantities of mesophilic bacteria, yeast, and molds considerably; nevertheless, it is unsuccessful at removing or even reducing mesophilic bacterial spores. Blanching has also been shown to lower lactic acid bacteria levels and total psychrotrophic bacteria counts [47]. Chemical composition variations take longer to notice, owing to the leaching of soluble elements that might impact protein levels [48]. Some proteins go through structural changes during the boiling process, including denaturation, crosslinking, and associations with lipids and carbohydrates. As a response, the proportion of hydrophilic sites available for water binding might well be reduced, and sorption properties may be changed, necessitating a more detailed protein study [49].

To minimize the quality loss and reduce the antimicrobial effects, each bug species must be treated with a blanching procedure. Blanching edible insects should lessen the microbiological dangers connected with their intake, and it might be augmented with ways to diminish bacterial spores quantity [50]. However, in order to suit the needs of customers and the processing industry, these treatments must preserve the nutritional value of the food as well as other crucial quality characteristics including texture and color.

4.2 Drying

The most extensively utilized technology for extending the shelf life of foods is drying. Traditional processes (such as roasting, frying, and sun-drying) are used in modern ways (for example, freeze-drying and microwave-assisted drying). Drying can lower total water content and, as a result, the amount of water available for metabolizing activities, such as enzymatic reactions and responses triggered by spoilage bacteria. Water action has a direct impact on microbial proliferation (aw). At aw 0.65, the vast majority of bacteria remain dormant. When aw is low, microorganisms slow down their development, but when water conditions improve, they can resume their progress [51, 52]. Reducing free water enhances dry matter concentration without harming tissues or the physical look of foods and is a key stage in the recovery of food products [53].

Solar, freeze-drying, and oven-drying are the popular techniques for drying entire edible insects, whereas freeze-drying, oven-drying, and nontraditional drying methods are mostly utilized for insect flours and powders. One of the chosen strategies for boosting human consumption of insects, mostly in Western countries, is drying and grinding whole, clearly recognized food insects into unrecognizable powders [54]. Drying extends the life of a product during distribution and storage.

Sun drying, one of the oldest drying methods, avoids microbiological contamination and even decreases or eliminates hazardous substances like neurotoxins; it also enhances the product's overall nutritional quality by inhibiting the enzyme protease inhibitors [55]. Furthermore, smoking is a curative and heating procedure that is regarded as one of the oldest methods for preserving all types of meat. The raw material is subjected to smoke created by wood pyrolysis. Insects are smoked in a dry environment, and a curing implementation is carried out concurrently with drying. The combined action of enzymes and heat promotes protein and lipid modifications during the process [56].

Irrespective of the blanching and drying techniques used on insects, they must be reheated prior to eating to remove any remaining bacteria. Total bacteria, Enterobacteriaceae, Staphylococcus, Bacillus, yeasts, and molds have all been reported to be eliminated by boiling dried insects for 30 minutes [47].

5. Production and processing of edible insect protein

Protein-enriched ingredients of edible insects have been obtained by using a procedure that has five basic steps which are:

- Initial treatment
- Defatting
- Solubilization and recovery of protein
- Purification of protein
- Drying

5.1 Initial treatment

The preliminary developmental stage is very contingent on the source product, in which entire insects are freeze-dried or roasted in the oven, pulverized, and mesh sieve to generate a fine powder that enhances protein solvent interaction in the subsequent steps [57].

5.2 Defatting

The form and substance of the precursor material have a big impact on the defatting process. For a few existing low-fat items, this step may simply be skipped [58]. The protein recovery and purity of the final product are directly affected by eliminating fat from the sample. Protein-lipid interactions restrict protein solubility due to the hydrophobic nature of lipids [59]. At the laboratory scale, the defatting procedure also increased insect protein output [60]. *T. molitor* and *Hermetia illucens*, for example, had their protein content increased from 57.8 ± 1.2 percent to 64.6 ± 0.3 percent and 34.7 ± 0.2 percent to 44.9 ± 1.4 percent, accordingly, due to hexane defatting process [61]. Defatting is an important step in making insect-based protein-enriched foods, given these findings and the high-fat content of most edible insects [62].

5.3 Solubilization and recovery of protein

To adjust protein solubility, the pH and ionic strength of the solubilization media are frequently changed to extract proteins from either constituent of defatted wheat. Because most food materials, including legumes and insects, have an isoelectric point (pI) of between 4 and 5, excessive pH values cause the protein surface to become charged, enhancing both electrostatic repulsion between proteins and protein-water associations [63–65]. To increase protein solubility, ionic strength can also be changed. Salt ions operate as a "shield" for proteins with opposing surface charges at low ionic strength (0.5–1 M), enhancing protein-water association and water solubility [66]. This method is known as salting in. The protein solution is normally centrifuged or gently eluted to complete the solubilization process. Proteins and other water-soluble constituents are retrieved from the supernatant, whereas water-insoluble polysaccharides and other minor components are eliminated from the pellet [54].

5.4 Purification of protein

Purifying specific proteins from an isolate or concentrate using chromatographic techniques is a helpful technique. Ion exchange, reversed-phase, and affinity chromatography rely on the variable affinity of proteins for the immobile stage [62, 67], and gel filtration chromatography isolates proteins by molecular weight [63]. Although chromatographic purification technologies have yet to be utilized on insects for food. For example, ion exchange chromatography was used to partially purify arginine kinase from yellow mealworms [68].

5.5 Drying

Drying can be used as the first step in the creation of insect flour and even the final step in the production of protein concentrates or isolates. Oven drying, fluidized bed drying, and microwave drying, for example, can all be utilized to dry solid samples and are thus utilized to make flour from whole insects prior to grinding. Depending on the study, the experimental setup and parameters may vary greatly. For example, it was found that oven-drying parameters for various insects ranged from 40 to 80°C for 7 to 48 hours, or till consistent weight was achieved [69]. Microwaves were used to dry *T. molitor* or *H. illucens* in a few articles comparing various drying processes. Depending on the experimental setup, microwave drying parameters ranged from 0.5 to 2 kW for 30 seconds or up to 20 minutes [70]. Fluidized bed drying, however less widespread, has been recorded as an alternate drying method for insect flour manufacture [18].

6. Contribution of insects to food and feed security

The contribution of edible insects in augmenting food supply is unpredictable. There is an extreme need to increase the traditional animal feed such as soy, grains, and fishmeal by using alternative sources. It is estimated that by 2030, over nine billion people and billions of animals will need to be fed. Furthermore, massive livestock production can cause land and water pollution and their overgrazing can also cause forest degradation that imparts a negative impact on our climate and damage our environment more rapidly. In order to reduce this rapidly destroying environment, alternative ways to feed livestock must be explored in which the consumption of edible insect provide innovation. For addressing food and feed security, insect farming is always the best option. Insects are ubiquitous, have high and rapid growth rates, and impart less environmental imprint throughout their lifecycle. They have high nutritional value, can be raised on a variety of different waste streams, and do not require any specific land for cultivation. In the coming decade, the usage of insects as poultry and aquaculture fodder is projected to be more common.

The usage of insects for feeding purpose is beneficial for us from health, environmental, and social perspective [71] which are described as follow:

6.1 Health benefits

Insect nutrition is determined by their life stage (metamorphic stage), habitat, and feed. It is commonly agreed, however, that:

- As insects possess high-quality protein and have high nutritional value so they are extremely useful for undernourished children as a food supplement. It has been indicated that insects have high fat content and they are also abundant in micronutrients.

- Insects have a lesser tendency of spreading zoonotic diseases, such as H1N1 (bird flu) and BSE (bovine spongiform encephalopathy) (mad cow disease).

6.2 Environmental benefits

- Because insects are cold-blooded, they have high feed to meal conversion rates. The estimated feed to meal conversion rate of insects is 2:1 as compared to calves whose estimated conversion ratio is 8:1.

- Most insects are likely to produce fewer greenhouse gases than typical animals. Pigs, for example, create 10–100 times the amount of greenhouse emissions per kilogram of body weight as mealworms.
- Insect farming requires less area than traditional animal husbandry.

6.3 Social benefits

- Harvesting and rearing technology is inexpensive, thus collecting insects and growing them require little technical or capital investment.
- Gathering insects from the wild, cultivating them, and selling them as street food can be a source of income for the lowest people of society.
- In industrialized, transitional, and developing economies, insect gathering, and farming can provide entrepreneurship prospects.
- Insects are reasonably easy to process for food and feed. Some species can be eaten in their entirety. Insects can also be ground into meals or made into pastes, and their proteins recovered.

7. Challenges and perspectives in insect food

One of the highly efficient protein sources is considered to be edible insects, which provide a chief source of protein for the future. The biggest challenge for consuming insect food is the neophobia that is associated with insect food [72]. Besides these, various other challenges are also present in consuming insects that have discussed below.

7.1 Aversion to insect consumption

Neophobia and disgust associated with trying insect is the biggest challenge to their integration into the food industry [73]. Studies showed that the two major barriers to consuming insects are neophobia and disgust, however, disgust plays a major role in comparison to neophobia [74]. In order to resolve this, scientists suggested it is more beneficial to enhance the quality, taste, and appearance of food.

7.2 Processing

Harvestation of insects has been done on a larger scale primarily for human consumption or pet and livestock feed [75]. As the utilization of insects has been increasing day by day so it is necessary to use more advanced and efficient processing methods to extract protein. Information about processing methods for yielding pure insect protein remains scarce and mainly performed at the laboratory level [76]. However, the purity and the quantity of insect protein solely depend on insect species, processing techniques, and rearing methods used. This future study emphasizes for extracting more insect protein by using optimal processes.

7.3 Allergenicity

Another main issue associated with edible insects is allergy. Anaphylactic reactions have been found in several studies after eating edible insects; these reactions may have been triggered by the existing link between insects and arthropods (such as shrimps and dust mites). Purines are made up of hydrogen, carbon, and nitrogen, as well as adenine and guanine, and are found in proteins. Uric acid is just another product of purine metabolism in humans that can be hazardous to gout sufferers. Undeniably cross allergic reactions to crustaceans and house dust mite protein have been associated with various insects. Co sensitivity and cross-reactivity between crustaceans, house dust mite, and edible insects have been studied and it was found that tropomyosin and arginine kinase are the major proteins responsible for allergic reactions [77]. Broekman *et al.* studied allergic reactions corresponding to insect consumption [78]. Different methods such as enzymatic hydrolysis and thermal treatment have been used to reduce the intensity of various allergens in various studies that are somehow effective [21].

7.4 Microbial hazards

One of the most serious challenges is the microbial hazards that are associated with edible insects. It has been found that total account of mesophilic aerobes is very high approx. to 7–8 Log10 CFU per gram in some edible insects parallel to that present in vegetables and fresh herbs [22]. Moreover, another serious hygienic concern is Enterobacteriaceae and found at the level of 106–1012 CFU/g [79], which is extremely high. To decrease the level, scientists suggest that prior to eating insect have been fasted for a day or it also greatly depend on food processing and rearing conditions [76]. Furthermore, edible insects might also possess a wide range of pathogens and spoilage bacteria that can be reduced by the processing step.

7.5 Environment

In using edible insects as a source of novel protein studies have been conducted by many scientists from an environmental point of view as environmental pollution is a serious issue [80]. Additionally, life cycle assessment of different insects provides lower environmental impact as compared to broiler but shows high environmental effect in contrast to plants [81].

7.6 Religious view about entomophagy

Food practices are influenced by culture and religious beliefs. The practice of eating insects is cited throughout religious literature in the Christian, Jewish and Islamic faiths. The consumption of insects depends on religion and ethnicity. It is a taboo among Muslims [82]. In the sacred books of the Christian, Jewish, and Islamic religions, there are different fragments related to entomophagy. Entomophagy in ancient times is a noted in the New Testament of the Bible with reference to John the Baptist, who appeared totally depended for his food on locusts and honey. In the Holy Bible use of locusts, crickets, and grasshoppers as food is mentioned as well as crickets and grasshoppers (Leviticus 11:22). In Islamic traditions, evidence of the use of insects as food extends to locusts and bees. In Judaism four species of locusts are recognized as kosher; that is, permitted for use as food [12, 83]. Entomophagy is also present in Jewish literature. Eating certain species of kosher locusts was largely accepted in ancient times.

7.7 Future directions

People should be educated about the importance of insects. The proper industry should be developed. A comparative analysis of the nutritional value of insects and red meat found that, in general, the value in regard to protein content, essential amino acids, and unsaturated fatty acids was similar [84]. People should have awareness that insects may prevent micronutrient deficiencies and play a role in the management of chronic diseases, such as cardiovascular disease, diabetes, and cancer [85].

Author details

Mahnoor Pervez* and Farkhanda Manzoor
Department of Zoology, Lahore College for Women University, Lahore, Pakistan

*Address all correspondence to: mahnoorentomology@gmail.com

References

[1] Per P-A. Food security: Definition and measurement. Food Security. 2009, 2009;**1**:5-7

[2] Timmer CP. The macro dimensions of food security: Economic growth, equitable distribution, and food Price stability. Food Policy. 2000;**25**:28

[3] FAO (Food and Agriculture Organization). The Sate of Food Insecurity in the World 2001. Rome: Food and Agriculture Organization of the United Nations; 2002

[4] Godfray HCJ, Beddington JR, Crute IR, Haddad L, Lawrence D, Muir JF, et al. Food security: The challenge of feeding 9 billion people sciences. Science. 2010;**327**(5967):812-818

[5] Westra S, Alexander L, Zwiers F. Global increasing trends in annual maximum daily precipitation. Journal of Climate. 2012;**23**:3904-3918

[6] Schlenker W, Roberts M. Nonlinear temperature effects indicate severe damages to U.S. crop yields under climate change. Proceedings of the National Academy of Sciences. 2009;**106**(37):15594-15598

[7] The Royal Society. Ground-Level Ozone in the 21st Century: Future Trends, Impacts and Policy Implications. Science Policy Report 15/08. London: The Royal Society Science Policy; 2008

[8] Tamburino L, Bravo G, Clough Y, Nicholas KA. From population to production: 50 years of scientific literature on how to feed the world. Global Food Security Index. 2020;**24**:100346

[9] Ramos-Elorduy J, Gonzalez EA, Hernandez AR, Pino JM. Use of *Tenebrio molitor* (Coleoptera: Tenebrionidae) to recycle organic wastes and as feed for broiler chickens. Journal of Economic Entomology. 2002;**95**(1):214-220

[10] Rumpold BA, Schluter OK. Nutritional composition and safety aspects of edible insects. Molecular Nutrition and Food Research. 2013;**57**(5):802-823

[11] Oonincx DG, van Itterbeeck J, Heetkamp MJ, van den Brand H, van Loon JJ, van Huis A. An exploration on greenhouse gas and ammonia production by insect species suitable for animal or human consumption. PLoS One. 2010;**5**(12):e14445

[12] Van Huis A. Potential of insects as food and feed in assuring food security. Annual Review of Entomology. 2013;**58**(1):563-583

[13] Raubenheimer D, Rothman JM. Nutritional ecology of entomophagy in humans and other primates. Annual Review of Entomology. 2013;**58**:141-160

[14] Guine FPR, Correia PR, Coelho., Costa, C. A. The role of edible insects to mitigate challenges for sustainability. Open Agriculture. 2021;**6**(1):24-36

[15] Beatriz A, Acosta E, Reyes A, Cristina MR, Dolores R, Celeste CIH. Benefits and challenges in the incorporation of insects in food products. Frontiers in Nutrition. 2021;**8**:687712

[16] Arrese EL, Soulages JL. Insect fat body: Energy, metabolism, and regulation. Annual Review of Entomology. 2010;**2010**

[17] Kourimska L, Adamkova A. Nutritional and sensory quality of edible insects. NFS Journal. 2016;**4**:22-26

[18] Kroncke N, Böschen V, Woyzichovski J, Demtroder S, Benning R. Comparison of suitable drying processes for mealworms (*Tenebrio molitor*). Innovative Food Science & Emerging Technologies. 2018;**50**:20-25

[19] Kulma M, Plachy V, Kourimska L, Vrabec V, Bubova T, Adamkova A, et al. Nutritional value of three Blattodea species used as feed for animals. Journal of Animal and Feed Sciences. 2016;**25**(4):354-360

[20] Belluco S, Losasso C, Maggioletti M, Alonzi C, Ricci A, Paoletti MG. Edible insects: A food security solution or a food safety concern? Animal Frontiers. 2015;**5**(2):25-30

[21] Pali-Scholl I, Meinlschmidt P, Larenas-Linnemann D, Purschke B, Hofstetter G, Rodriguez-Monroy FA, et al. Edible insects: Cross-recognition of IgE from crustacean- and house dust mite allergic patients, and reduction of allergenicity by food processing. World Allergy Organization Journal. 2019, 2019;**12**(1):100006

[22] Fasolato L, Cardazzo B, Carraro L, Fontana F, Novelli E, Balzan S. Edible processed insects from e-commerce: Food safety with a focus on the Bacillus cereus group. Food Microbiology. 2018;**76**:296-303

[23] Bukkens SGF. The nutritional value of edible insects. Ecology of Food and Nutrition. 1997;**36**(2-4):287-319

[24] Sales-Campos H, Reis de Souza P, Crema Peghini B, Santana da Silva J, Ribeiro Cardoso C. An overview of the modulatory effects of oleic acid in health and disease. Mini Reviews in Medicinal Chemistry. 2013, 2013;**13**(2):201-210

[25] Finke MD. Nutrient composition of bee brood and its potential as human food. Ecology of Food and Nutrition. 2005;**44**(4):257-270

[26] Bukkens SG, Paoletti MG. Insects in the human diet: Nutritional aspects. In: Ecological Implications of Minilivestock; Role of Rodents, Frogs, Snails, and Insects for Sustainable Development. Enfield, USA: Science Publishers, Inc; 2005. pp. 545-577

[27] Tang C, Yang D, Liao H, Sun H, Liu C, Wei L, et al. Edible insects as a food source: A review. Food Production, Processing and Nutrition. 2019;**8**(1):1-8

[28] Jozefiak D, Jozefiak A, Kieronczyk B, Rawski M, Swiatkiewicz S, Dlugosz J. Insects - a natural nutrient source for poultry - a review. Annals of Animal Science. 2016;**16**(2):297-313

[29] Krishnan R, Sherin L, Muthuswami M, Balagopal R, Jayanthi C. Seri waste as feed substitute for broiler production. Sericologia. 2011;**51**(3):369-383

[30] Awoniyi TAM, Adetuyi FC, Akinyosoye FA. Microbiological investigation of maggot meal, stored for use as livestock feed component. Journal of Food, Agriculture and Environment. 2004;**2**(3-4):104-110

[31] Liu Y-S, Wang F-B, Cui J-X, Zhang L. Recent status and advances on study and utilization of Tenebrio molitor. Journal of. Environmental Entomology. 2010;**32**:106-114

[32] Ramos-Elourdy J, Pino-Moreno JM, Martinez-Camacho VH. Could grasshoppers be a nutritive meal. Food and Nutrition Sciences. 2012;**3**:164-175

[33] Gahukar RT. Entomophagy and human food security. International Journal of Tropical Insect Science. 2011;**31**(3):129-144

[34] Shelomi M. Why we still don't eat insects: Assessing entomophagy promotion through a diffusion of innovations framework. Trends in Food Science & Technology. 2015;**45**(2):311-318

[35] Raheem D, Carrascosa C, Oluwole OB, Nieuwland M, Saraiva A, Millan R, et al. Traditional consumption of and rearing edible insects in Africa, Asia and Europe. Critical Reviews in Food Science and Nutrition. 2018;**59**(14):1-20

[36] Van Huis A, Oonincx DGAB. The environmental sustainability of insects as food and feed. A review. Agronomy for Sustainable Development. 2017;**37**(5):43

[37] Durst PB, Hanboonsong Y. Small-scale production of edible insects for enhanced food security and rural livelihoods: Experience from Thailand and Lao People's Democratic Republic. Journal of Insects as Food and Feed. 2015;**1**(1):25-31

[38] Cerritos R, Cano-Santana Z. Harvesting grasshoppers *Sphenarium purpurascens* in Mexico for human consumption: A comparison with insecticidal control for managing pest outbreaks. Crop Protection. 2008;**27**(3):473-480

[39] Lara-Juarez P, Rivera JRA, Lara PC, Reyes-Aguero JA. Collecting pupae (escamoles) of *Liometopum apiculatum* (Hymenoptera, Formicidae, Dolichoderinae) in the San Luis Potosi high plain, Mexico. Interciencia. 2018;**43**(11):763-769

[40] Payne CLR, Van Itterbeeck J. Ecosystem services from edible insects in agricultural systems: A review. Insects. 2017;**8**(1):E24

[41] Cerritos-Flores R, Ponce-Reyes R, Rojas-Garcia F. Exploiting a pest insect species *Sphenarium purpurascens* for human consumption: Ecological, social, and economic repercussions. Journal of Insects as Food and Feed. 2014, 2014;**1**(1):75-84

[42] Gahukar, R. T. Edible insects farming: Efficiency and impact on family livelihood, food security, and environment compared with livestock and crops. 2016;**7**:85-111

[43] Varelas V, Langton M. Forest biomass waste as a potential innovative source for rearing edible insects for food and feed – A review. Innovative Food Science & Emerging Technologies. 2017;**41**:193-205

[44] Van Thielen L, Vermuyten S, Storms B, Rumpold BA, Van Campenhout L. Consumer acceptance of foods containing edible insects in Belgium two years after their introduction to the market. Journal of Insects as Food and Feed. 2018;**5**(1):35-44

[45] Xiao H-W, Bai J-W, Sun D-W, Gao Z-J. The application of superheated steam impingement blanching (SSIB) in agricultural products processing—A review. Journal of Food Engineering. 2014;**132**:39-47

[46] Wynants E, Crauwels S, Verreth C, Gianotten N, Lievens B, Claes J, et al. Microbial dynamics during production of lesser mealworms (*Alphitobius diaperinus*) for human consumption at industrial scale. Food Microbiology. 2018;**70**:181-191

[47] Vandeweyer D, Lenaerts S, Callens A, Van Campenhout L. Effect of blanching followed by refrigerated storage or industrial microwave drying on the microbial load of yellow mealworm larvae. Food Control. 2017;**71**:311-314

[48] Niamnuy C, Devahastin S, Soponronnarit S. Changes in protein

compositions and their effects on physical changes of shrimp during boiling in salt solution. Food Chemistry. 2008;**108**(1):165-175

[49] Derossi A, Severini C. Understanding the drying kinetic and hygroscopic behaviour of larvae of yellow mealworm (*Tenebrio molitor*) and the effects on their quality. Journal of Insects as Food and Feed. 2016;**2**(4):233-243

[50] Caparros Megido R, Desmedt S, Blecker C, Bera F, Haubruge E, Alabi T, et al. Microbiological load of edible insects found in Belgium. Insects. 2017;**8**(1):12

[51] Grabowski NT, Klein G. Microbiology of cooked and dried edible Mediterranean field crickets (*Gryllus bimaculatus*) and superworms (*Zophobas atratus*) submitted to four different heating treatments. Food Science and Technology International. 2016;**23**(1):17-23

[52] Grabowski NT, Klein G. Microbiology of processed edible insect products—Results of a preliminary survey. International Journal of Food Microbiology. 2017;**243**:103-107

[53] Lamidi RO, Jiang L, Pathare PB, Wang YD, Roskilly AP. Recent advances in sustainable drying of agricultural produce: A review. Applied Energy. 2019;**233-234**:367-385

[54] Menozzi D, Sogari G, Veneziani M, Simoni E, Mora C. Eating novel foods: An application of the theory of planned behaviour to predict the consumption of an insect-based product. Food Quality and Preference. 2017;**59**:27-34

[55] Niassy S, Fiaboe KKM, Affognon HD, Akutse KS, Tanga MC, Ekesi S. African indigenous knowledge on edible insects to guide research and policy. Journal of Insects as Food and Feed. 2016;**2**(3):161-170

[56] Tiencheu B, Womeni HM, Linder M, Mbiapo FT, Villeneuve P, Fanni J, et al. Changes of lipids in insect (*Rhynchophorus phoenicis*) during cooking and storage. European Journal of Lipid Science and Technology. 2013;**115**(2):186-195

[57] Zielinska E, Karas M, Baraniak B. Comparison of functional properties of edible insects and protein preparations thereof. A review of literature. Lebensmittel-Wissenschaft und-Technologie-Food Science and Technology. 2018;**91**:168-174

[58] Boye J, Zare F, Pletch A. Pulse proteins: Processing, characterization, functional properties and applications in food and feed. Food Research International. 2010;**43**(2):414-431

[59] Azagoh C, Ducept F, Garcia R, Rakotozafy L, Cuvelier ME, Keller S. Mezdour,.Extractionandphysicochemical characterization of Tenebrio molitor proteins. Food Research International. 2016;**88**(Pt A):24-31

[60] Choi BD, Wong NAK, Auh JH. Defatting and sonication enhances protein extraction from edible insects. Korean Journal for Food Science of Animal Resources. 2017;**37**(6):955-961

[61] Bubler S, Rumpold BA, Jander E, Rawel HM, Schluter OK. Recovery and techno-functionality of flours and proteins from two edible insect species: Meal worm (*Tenebrio molitor*) and black soldier fly (*Hermetia illucens*) larvae. Heliyon. 2016;**2**(12):e00218

[62] Tzompa-Sosa DA, Yi L, van Valenberg HJF, van Boekel MAJS, Lakemond C, n M. M. Insect lipid profile: Aqueous versus organic solvent-based

extraction methods. Food Research International. 2014;**62**:1087-1094

[63] Boye J, Barbana C. Protein processing in food and bioproduct manufacturingand techniques for analysis. In: Dunford NT, editor. Food and Industrial Bioproducts and Bioprocessing. USA: John Wiley & Sons, Ltd., Publication; 2012. pp. 85-113

[64] Lam ACY, Can Karaca A, Tyler RT, Nickerson MT. Pea protein isolates: Structure, extraction, and functionality. Food Reviews International. 2018;**34**(2):126-147

[65] Singhal A, Karaca AC, Tyler R, Nickerson M. Pulse proteins: From processing to structure-function relationships. Grain Legumes. 2016;**2016**(8):56-78

[66] Duong-Ly KC, Gabelli SB. Chapter seven - salting out of proteins using ammonium sulfate precipitation. In: Lorsch J, editor. Methods in Enzymology. Vol. 541. USA: Elsevier Inc; 2014. pp. 85-94

[67] Urh M, Simpson D, Zhao K. Chapter 26 affinity chromatography: General methods. In: Burgess RR, Deutscher MP, editors. Methods in Enzymology. Vol. 463. USA: Elsevier Inc; 2009. pp. 417-438

[68] Francis F, Doyen V, Debaugnies F, Mazzucchelli G, Caparros R, Alabi T, et al. Limited cross reactivity among arginine kinase allergens from mealworm and cricket edible insects. Food Chemistry. 2019;**276**:714-718

[69] Melgar-Lalanne G, Hernandez-Alvarez AJ, Salinas-Castro A. Edible insects processing: Traditional and innovative technologies. Comprehensive Reviews in Food Science and Food Safety. 2019;**18**(4):1166-1191

[70] Huang C, Feng W, Xiong J, Wang T, Wang W, Wang C, et al. Impact of drying method on the nutritional value of the edible insect protein from black soldier fly (*Hermetia illucens L.*) larvae: Amino acid composition, nutritional value evaluation, in vitro digestibility, and thermal properties. European Food Research and Technology. 2018;**245**(1):11-21

[71] Kuyper E, Vitta B, Dewey K. Novel and underused food sources of key nutrients for complementary feeding. A&T Tech Brief. 2013;**6**:1-8

[72] Jensen NH, Lieberoth A. We will eat disgusting foods together – Evidence of the normative basis of Western entomophagy-disgust from an insect tasting. Food Quality and Preference. 2019;**72**:109-115

[73] Verbeke W. Profiling consumers who are ready to adopt insects as a meat substitute in a Western society. Food Quality and Preference. 2015;**39**: 147-155

[74] La Barbera F, Verneau F, Amato M, Grunert K. Understanding westerners' disgust for the eating of insects: The role of food neophobia and implicit associations. Food Quality and Preference. 2018;**64**:120-125

[75] Van Huis A, van Itterbeeck J, Klunder H, Mertens E, Halloran A, Muir G, et al. Edible Insects - Future Prospects for Food and Feed Security. FAO Forestry Paper. Vol. 1712013. Rome: Food and Agriculture Organization of the United State

[76] Zhao X, Vazquez-Gutierrez JL, Johansson DP, Landberg R, Langton M. Yellow mealworm protein for food purposes - extraction and functional properties. PLoS One. 2016;**11**(2):e0147791

[77] Ribeiro JC, Cunha LM, Sousa-Pinto B, Fonseca J. Allergic risks of consuming edible insects: A systematic review. Molecular Nutrition & Food Research. 2018;**62**(1):1-12

[78] Broekman H, Verhoeckx KC, den Hartog Jager CF, Kruizinga AG, Pronk-Kleinjan M, Remington BC, et al. Majority of shrimp-allergic patients are allergic to mealworm. The Journal of Allergy and Clinical Immunology. 2016;**137**(4):1261-1263

[79] Fernandez-Cassi X, Supenu A, Jansson A, Boqvist S, Vagsholm I. Novel foods: A risk profile for the house cricket (*Acheta domesticus*). EFSA Journal. 2018;**5**:137-157

[80] Oonincx DG, van Broekhoven S, van Huis A, van Loon JJ. Feed conversion, survival and development, and composition of four insect species on diets composed of food by-products. PLoS One. 2015;**10**(12):e0144601

[81] Smetana S, Schmitt E, Mathys A. Sustainable use of *Hermetia illucens* insect biomass for feed and food: Attributional and consequential life cycle assessment. Resources, Conservation and Recycling. 2019;**144**:285-296

[82] Djouffa KML, Saidou C, Ali A, Mohammadou BA. Tchiegang C. Edible insects in the Adamawa and eastern regions of Cameroon: Collection, processing and consumption. Journal of Insects as Food and Feed. 2021;7(12):483-494

[83] Govorushko S. Global status of insects as food and feed source: A review. Trends in Food Science and Technology. 2019;**91**:436-445

[84] Orkusz A. Edible insects versus meat-nutritional comparison: Knowledge of their composition is the key to good health. Nutrients. 2021;**13**(4):1207

[85] Nowakowski AC, Miller AC, Miller ME, Xiao H, Wu X. Potential health benefits of edible insects. Critical Reviews in Food Science and Nutrition. 2021;**2**:1-10

Chapter 4

Genetically Modified Food for Ensuring Food Security Issues

Sujay Kumar Bhajan, Mohammad Mahmood Hasan, Md. Anwarul Haque and Mohammad Nurul Islam

Abstract

Genetic modification is a technology that allows incorporating beneficial genes from other organisms. One of the major challenges of current era is feeding the consisting growing population of the world. Reduced cultivatable land and climate change have made it even more challenging. Through genetic engineering technology it is possible to develop high yielding, disease resistance and climate resilient crop varieties within shorter period of time than the conventional ways. Current global status of GM crop cultivation and production has already proved that GM crops have the potential to achieve food security for the future world. However, in most of the countries, public has misperception of its risks on human health and environment. Due to such misconception most of the countries have not adopted GM crops yet which could make it difficult to utilize GM crops to achieve food security. Therefore, in this book chapter we discussed on how genetically modified crops are developed, what beneficial traits are usually incorporated, how they are assessed for human health risks and regulations for selling genetically modified foods in the market. Such discussion would help common people to understand how this technology can help us ensuring food security of the world.

Keywords: genetically modified organisms (GMOs), genetically modified foods (GMF), food security, safety assessment, biosafety

1. Introduction

The growth of the human population will increase day by day and create significant challenges for agricultural production. Now a days food security is one of the major concerns for the growing global population. Nearly 870 million people suffer from malnutrition; most of them are living in the developing countries of Africa, Asia, and South America [1]. Additionally, climate change and environmental deterioration are currently reducing the available agricultural land, creating additional challenges to fill the increasing food demand [2]. The use of modern biotechnology, including genetic modification techniques, has been proposed as a way to reduce the environmental footprint, by improving food quality and increasing productivity [3]. Genetically Modified Food (GMF) means any food containing or derived from a genetically engineered organism [4]. The majority of the biotech crops available on the global market have been genetically manipulated to express one of these basic traits: resistance to insects or viruses, tolerance to certain herbicides and nutritionally enhanced quality. In 2019, the 24th year of commercialization of biotech crops, 190.4 million hectares of biotech crops were planted by up to 17 million farmers in 29 countries. From the initial planting of 1.7 million hectares in 1996 when the first biotech crop was commercialized, the 2019 planting indicates ~112-fold increase Thus, biotech crops are considered as the fastest adopted crop technology in the history of modern agriculture [5]. Major producers of GM crops include USA, Argentina, Canada, and China. In the US, about 80% of maize, cotton and soya are biotech varieties [6]. Several safety issues are considered for the release or commercialization of genetically modified crops such as safety issues for human health, toxicity analysis, allergenicity analysis and biosafety assessment of genetically modified foods. This book chapter discusses the overall process for developing of genetically modified crops, brief history of global adoption of genetically modified crops, main agricultural traits and current status of the approved genetically modified crops and its role in food Security.

2. The overall process for developing of genetically modified crops

The crop plants developed through modification of genetic composition using genetic engineering techniques are referred to as genetically modified (GM) crops, transgenic crops or genetically engineered (GE) crops. The overall process for developing of genetically modified crops illustrated in (**Figure 1**) [7]. Different steps are involved in the genetic transformation work like Identification and isolation of gene of interest (transgene), cloning into suitable plasmid vector and transformation of desired gene into the plant cells, confirmation of transformation by molecular techniques.

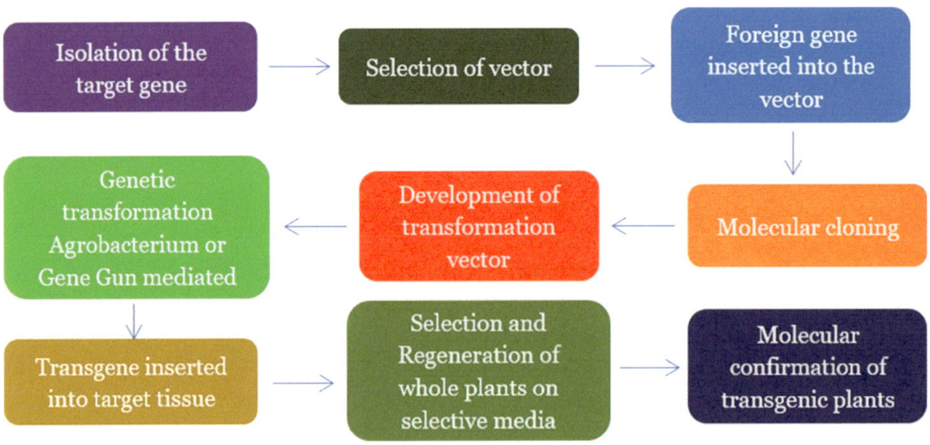

Figure 1.
The overall process for developing of genetically modified crops.

2.1 Identification and isolation of the target genes

To produce a GM plant, scientists first identify the desired trait missing in the cultivar, such as resistance to biotic and abiotic factors (salt, cold, drought, herbicides, pesticides, insects, nutritional quality etc.). Then, by using existing knowledge they find an organism (plant, animal, or microorganism) processing the desired trait. Finally, the desired genes are isolated using suitable molecular tools.

2.2 Molecular cloning of target gene and development of plant transformation vector

A proper gene construct is crucial for the success of producing ideal transgenic line. A vector is a DNA molecule that has the ability to replicate inside the host to which the desired gene has integrated for cloning. A vector acts as a vehicle that transports the gene of interest into a target cell for replication and expression [8]. Common vector consists of three components: an origin of replication, multiple cloning site or recombination site and selectable marker gene. The common vectors include-plasmids, cosmids, bacteriophages, bacterial artificial chromosome (BAC), yeast artificial chromosome (YAC), etc. which is used for the production of genetically modified crops [9]. The most commonly used plant transformation vectors are Ti plasmid binary vectors of *Agrobacterium tumefaciens.*

2.3 Transformation of desired gene into the plant cells

Transformation of transgenes into the plant cells can be carried out through indirect or direct gene transfer methods. An indirect gene transfer (vector-mediated) method involves the introduction of exogenous DNA into the plant genome via biological vectors, whereas direct gene transfer methods involve the introduction of exogenous DNA directly into plant genome through physical or chemical reactions. The important physical and chemical gene transfer methods are electroporation, microinjection, particle bombardment, Polyethylene glycol and Liposome mediated gene transfer. *Agrobacterium*-mediated transformation is the most common technique used in plant transformation as it is efficient and effective in a wide range of plants [10]. This method had successfully transformed a broad variety of plants such as rice, maize, barley, and tobacco. Among the physical methods particle bombardment biolistic is the most common one used in plant transformation [11]. Currently, transient and stable transformation can be achieved through the *Agrobacterium*-mediated and biolistic methods. In the *Agrobacterium*-mediated method, the T-DNA region is inserted into the plant genome forming a stable transformant, whereas the non-integrated T-DNA plasmid expresses the transgene transiently. In the biolistic method or other direct gene transfer methods such as electroporation, transient and stable expression of the transgenes are usually dependent on the plasmid or transgene constructs.

2.4 Confirmation of transgene integration by molecular techniques

Confirmation of transgene integration has to be done through an appropriate method based on the transgene constructs, selectable marker, and reporter gene used. The antibiotic resistance genes are screened by the addition of herbicides or antibiotics to the growing media to distinguish transformed plant cells from the non-transformed plant cells. However, this method requires a large quantity of antibiotics and herbicides that are expensive and worsen by the risk of horizontal gene transfer to other bacteria.

An alternative screening method such as polymerase chain reaction (PCR) and reporter gene expression are used for the identification of transgenic plants. Some reporter genes such as the GFP, GUS, CAT, LacZ and Luc are used for the expression of these genes could be observed visually or directly under microscopy [12]. The GUS expression can also be detected through histochemical assay in which the localization of the transgene can be observed. Southern blotting is a molecular method which is generally used to identify the number of transgenes (copy number) inserted into the host genome as well as for the detection of transgene integrity and transgene rearrangement [13]. The polymerase chain reaction (PCR) method is one of the most sensitive and easiest methods among all the molecular techniques employed for the verification of the transgene. The PCR is generally done with primers specific to the site of plasmid constructs and gene of interest used for development of the transgenic plants. Successful amplification of the DNA fragment with expected band indicates the possible presence of transgene, and this DNA fragment is further confirmed through DNA sequencing. A real-time PCR provides fast, sensitive, and high-throughput molecular PCR–based analysis compared to the traditional Southern blot analysis especially in the area of transgene copy number and zygosity detection in transgenic plants [14]. In recent years, the availability of NGS tools and bioinformatics resources facilitates the study of genome and molecular characterization of complex traits. Hence, NGS approach provides an alternative high-resolution analysis tool for transgenes insertion in GM crops [15].

2.5 Regeneration of the whole plant from the transformed cells

Expression of transgene and production of functional protein is important for development of genetically modified foods. Successful genetic engineering of plants often requires the regeneration of whole plants from transformed plant cells. Plant regeneration usually involves culturing plant tissue in the presence of specific plant hormones under sterile conditions.

3. Brief history of global adoption of genetically modified crops

The history of GM crop in agriculture started with Flavr Savr™ Tomato, the first ever GM crop approved for human consumption in 1994 by the Food and Drug administration, USA [16]. This GM tomato was developed by former Monsanto Company (currently Bayer Crop Science) with a longer shelf life. Later on, in 1995 Mexico and Canada adopted this GM tomato for consumption. Till then many countries started to adopt GM crops and according to ISAAA, in 2019, a total of 71 countries have adopted GM crops among which 29 countries directly cultivated different types of GM crops including cereal, vegetables, oilseed, fruits, and flowers, whereas rest of the countries imported GM food [17]. Most crops have been genetically modified for herbicide tolerance and insect resistance [18], however, several crops have been also targeted for modified quality, for example, modified flower color, modified nutrient quality, longer shelf life and so on [19].

4. Commercially approved GM crops and their commercial traits

The commercially approved GM crops around the world and their desired traits according to the reports of ISAAA are described below [17].

4.1 Cereal crops

Some important cereal crops such as maize, rice and wheat are approved for their commercial traits. Maize is the second most cultivated GM crop in the world and USA is the major cultivator and producer of GM corns. Maize has been genetically modified mainly for herbicide tolerance and insect resistance. Several commercial GM corns are stacked with the aforementioned traits, for instance, Agrisure® Duracade™ 5222 by Syngenta. Few are modified for better quality, such as, Enogen™, that contains a synthetic gene derived from *Thermococcales* spp.

Like maize, rice has also been genetically modified for herbicide tolerance and insect resistance. Liberty Link™ rice developed by Bayer Crop Science is a herbicide tolerance variety containing a gene (*bar*) from *Streptomyces hygroscopicus.* It was first approved in the USA in 2000 and later on other countries have also approved. Till date, 9 more countries including Canada, Argentina, Australia, Colombia, Honduras, Mexico, New Zealand, Philippines, Russia and South Africa have approved this GM rice. The commercially available insect tolerant GM rice is BT Shanyou 63 that was only approved by China in 2009. Another GM rice of similar trait is Huahui-1 that was approved in the USA in 2018. The most famous GM rice would be the Golden Rice. It is adopted for cultivation in Philippines in 2018, however, in 2017-2018, USA, Canada, Australia and New Zealand have also approved this as food and feed.

Roundup Ready™ wheat, a glyphosate herbicide resistant variety is the only GM wheat in the world developed by Monsanto Company. Initially, in 2004, 4 countries including Australia, Colombia, New Zealand, and the United States approved this for direct use and processing as food, however, Monsanto reportedly stopped pursuing the regulatory process and other countries also have withdrawn.

4.2 Oilseeds crops

Among the oilseed crops soybean, canola and sunflower are more important. Soybean is the most adopted and most produced GM crop in the whole world. The herbicide tolerant varieties Liberty Link® soybean and Contrivance were developed by BASF group. It is approved to be used as food and feed in more than 20 countries. Some other herbicide tolerant varieties are Liberty Link® GT27™, Roundup Ready™ soybean, Genuity® Roundup Ready™ 2 Xtend™, Genuity® Roundup Ready 2 Yield™, and Optimum GAT™. Apart from these, Conkesta Enlist E3™ Soybean and Intacta™ Roundup Ready™ 2 Pro, developed by Dow AgroSciences LLC and Monsanto Company, respectively, are stacked with herbicide tolerance and insect resistance capacity. The Vistive Gold™ developed by Monsanto Company and the Plenish™, developed by DuPont (Pioneer Hi-Bred International Inc.) are stacked with herbicide tolerance and enhanced oleic acid content. These are approved in more than 15 countries. Lastly, the Verdeca HB4 Soybean is an abiotic stress tolerant GM soybean developed by Verdeca. It is currently approved in 6 countries for food and feed purposes.

The genetic modification in Argentine Canola (*Brassica napus*) is more focused on herbicide tolerance. To name a few herbicides tolerant GM canola varieties would be Optimum® Gly canola, Roundup Ready™ Canola, Liberty Link™ Independence™, InVigor™ Canola, Navigator™ Canola, etc. Though there is no insect resistant GM canola available few have been genetically modified for improved quality. For example, Laurical™ Canola developed by Monsanto Company has increased levels of a triglyceride, lauric acid. However, it is only adopted for cultivation in the USA and Canada. There is only one GM Polish Canola (*Brassica rapa*) is available named Hysyn 101 RR Roundup-Ready™. It was developed by the University of Florida and only adopted for cultivation and approved for use in Canada (1997). GM safflower is adopted for cultivation only in Australia and it was approved in 2018 for food and feed purposes. This has a down regulated expression of *fatB* and *fad2.2* genes for lowering the fat content.

4.3 Vegetables and fruits

Commercially important vegetables and fruits are potato, tomato, eggplant, common bean, squash, apple, papaya, plum and pineapple. Primary potato trait that was focused for GM potatoes is modified product quality, such as, Starch potato, Amflora™, for reduced amylose and increased amylopectin content; Innate®, Cultivate, Innate® Generate, Innate® Accelerate, and Innate® Invigorate for lower reducing sugar. Several modified product quality GM potatoes are stacked with disease resistance capacity against blight (Innate® Acclimate, Simplot Innate) and potato virus Y (Hi-Lite NewLeaf™ Y potato). Commercially available coleopteran insect resistant GM potatoes are Superior NewLeaf™ potato, Elizaveta plus, Lugovskoi plus, Atlantic NewLeaf™ potato, etc.

According to ISAAA, 11 GM tomatoes have been developed so far for commercial purposes and 9 of them are for a single trait, that is, delayed fruit ripening or delayed fruit softening for longer shelf life. FLAVR SAVR™ by Monsanto Company and three others developed by Zeneca Plant Science and Petoseed Company followed the same strategy for delayed fruit softening. These four are only cultivated in the USA, however, Canada and Mexico have approved these for food purposes around 1995–1996. Some other GM varieties were developed by targeting the genes that are responsible for ethylene production, the fruit ripening hormone. This one was adopted and approved as food and feed in the USA in 1996. China approved 2 GM tomato varieties for cultivation and domestic uses with delayed ripening in 1997 and 1999 that was developed by the Huazhong Agricultural University (China) and Institute of Microbiology, CAS (China), respectively. The DNA Plant Technology Corporation (USA) developed another GM tomato by reducing the expression of *ACC synthase* gene resulting in reduced ethylene production, thus, delayed fruit ripening. Monsanto Company developed another one with resistance capacity against lepidopteran insects using *cry1Ac* gene and was approved in 1998 and in 2000 for Canada.

The only available GM eggplant is BARI bt Begun developed by Maharashtra Hybrid Seed Company (MAHYCO). It mainly contains *cry1Ac* gene from *Bacillus thuringiensis* subsp. Kurstaki strain HD73 that confers resistance against lepidopteran insects by selectively damaging their midgut lining. Bangladesh is the only country that approved the cultivation of this GM eggplant in 2013. Recently, Philippines approved it for food and feed uses but did not adopt for cultivation. In Bangladesh it is only approved for direct use or processing as food only.

BRS FC401 RMD is the only commercially available GM common bean (*Phaseolus vulgaris*). It was developed by EMBRAPA (Brazil) and is resistant against Bean Golden Mosaic Virus (BGMV). It inhibits the replication of viral protein. Brazil is the only country to cultivate and use this GM crop as food and feed. Seminis Vegetable Seeds (Canada) and Monsanto Company (Asgrow) developed the GM squash that is stacked with resistance capacity against three plant viruses, e.g., *Cucumber mosaic cucumovirus* (CMV), *Zucchini yellow mosaic* Potyvirus (ZYMV) and *Watermelon mosaic potyvirus 2* (WMV2). Cultivation of this variety was stared in 1996 in the USA and approved for food and feed usage in 1997. No other country has approved this one yet. Canada approved it food use in 1998.

Fruits.

Apple: Arctic™, Arctic™ "Golden Delicious" Apple, and Arctic™ Fuji Apple, all these 3 GM apple varieties has been developed for non-browning phenotype by Okanagan Specialty Fruits Incorporated. Arctic™ Fuji Apple was first adopted in Canada in 2018 and then in the USA in 2019. The other 2 are also approved only in these 2 countries and their cultivation started in 2015.

Papaya ringspot virus is the greatest concern of papaya cultivation. Therefore, all the GM papayas are aimed to disease resistance trait. Pathogen derived resistance mechanism is used in these GM papayas to check the viral infection. The USA and Japan started cultivation in 1996 and 2011, respectively. Canada only approved its use for food purposes in 2003. Another commercially available one is Huanong No. 1 developed by South China Agricultural University for the same disease resistance purpose. China is the only country that started its cultivation in 2006.

The only GM pineapple is Rosé, developed by Del Monte Fresh Produce Company. It has traits of delayed ripening, modified fruit color, and increased level of lycopene and beta-carotene. It is approved only in the USA (2016) and Canada (2021).

4.4 Sugar crops

Brazil and Indonesia are the only 2 countries to cultivate GM sugarcane. Brazil started the cultivation insect resistant GM sugarcane and approved its usage for food and feed purposes from 2017. It was developed by Centro de Tecnologia Canavieira (CTC) that contains *cry1Ab* gene for conferring resistance against lepidopteran insects. Canada and United States have approved its usage as food. On the other hand, Indonesia adopted the cultivation of drought resistant GM sugarcane and approved to be used as food and processing food products from 2011. It contains *EcBetA* gene catalyzes the production of the osmoprotectant compound glycine betaine conferring tolerance to water stress. Only 3 countries adopted the cultivation of GM sugar beet, Roundup Ready™ sugar beet. Australia and the USA approved in 2005 and Japan in 2007. However, Australia, Canada, China, Colombia, European Union, Japan, Mexico, New Zealand, Philippines, Russia, Singapore, South Korea, Taiwan and USA have approved its use as direct food or processed food. This GM sugar beet was developed by Monsanto Company. There are two more commercially available GM sugar beet plants are present, they are InVigor™ sugar beet developed by Novartis Seeds and Monsanto Company and Liberty Link™ sugar beet developed by Bayer CropSciene.

4.5 Others

Monsanto Company and Forage Genetics International have jointly developed GM alfalfa plants for herbicide tolerance (Roundup Ready™ Alfalfa) and reduced content of guaiacyl (G) lignin (HarvXtra™). These are approved in Australia, Canada, Japan, Mexico, New Zealand, Philippines, Singapore, South Korea and the United States of America in the years ranging from 2004 to 2008. Sweet peppers Approved for cultivation in China in 1998, developed by Beijing University. The GM sweet pepper confers resistance against the Cucumber Mosaic Virus (CMV) through pathogen derived resistance mechanism. Tobacco has been genetically modified for cultivation purposes targeting 2 commercial traits. First one by SEITA S.A. (France) for oxynil herbicide tolerance and the second one for reduce nicotine content developed by Vector Tobacco

Inc. (USA). The first one was approved in 1994 for the European Union and the latter one in 2002 for the United States.

5. Worldwide production and consumption of genetically modified foods

In 2019, According to ISAAA, 29 countries cultivated GM crops among which the USA is the largest producer of the GM crops covering 37.9% of the total GM crop area in the world [5]. USA has adopted the highest number of biotech crops (**Table 1**). Canada is another major GM crop grower that has reached 95% adoption rate of GM canola. In Latin America, 10 countries cultivate biotech crops, they are, Mexico, Costa Rica, Honduras, Columbia, Chile, Bolivia, Argentina, Uruguay, Paraguay, and Brazil. Brazil is the top biotech crop cultivator developing country. In Europe, only two countries adopted the cultivation of GM crop, Spain, and Portugal. These 2 countries only cultivate GM Maize. Whereas, in Africa, most of the countries cultivate GM cotton, like, Nigeria, Sudan, Ethiopia, Malawi, and Eswatini. South Africa cultivates three different GM crops, they are soybean, maize, and cotton. In Asia, Pakistan, India, Myanmar cultivates GM cotton. Bangladesh is the only country to cultivate BT Brinjal in the whole world. Indonesia only cultivates GM sugarcane, whereas Vietnam and Philippines are only focused on the cultivation of GM maize. China cultivates GM papaya and GM cotton. In Oceania, Australia is the only country to cultivate GM crops, like GM cotton, GM alfalfa and GM safflower. Australia is the exclusive grower

Country	Total area cultivated (Million hectares)	GM crops
USA	71.5	Maize, soybeans, cotton, alfalfa, canola, sugar beets, potatoes, papaya, squash, apples
Brazil	52.8	Soybeans, maize, cotton, sugarcane
Argentina	24.0	Soybeans, maize, cotton, alfalfa
Canada	12.5	Canola, soybeans, maize, sugar beets, alfalfa, potatoes
India	11.9	Cotton
Paraguay	4.1	Soybeans, maize, cotton
China	3.2	Cotton, papaya
South Africa	2.7	Maize, soybeans, cotton
Pakistan	2.5	Cotton
Bolivia	1.4	Soybeans
Uruguay	1.2	Soybeans, maize
Philippines	0.9	Maize
Australia	0.6	Cotton, canola, safflower
Myanmar	0.3	Cotton
Sudan	0.2	Cotton
Mexico	0.2	Cotton
Spain	0.1	Maize
Colombia	0.1	Cotton, maize

Country	Total area cultivated (Million hectares)	GM crops
Vietnam	0.1	Maize
Honduras	<0.0.5	Maize
Chile	<0.05	Maize, canola
Malawi	<0.05	Cotton
Portugal	<0.05	Maize
Indonesia	<0.05	Sugarcane
Bangladesh	<0.05	Brinjal
Nigeria	<0.05	Cotton
Eswatini	<0.05	Cotton
Ethiopia	<0.05	Cotton
Costa Rica	<0.05	Cotton, pineapple
Total	**190.4**	

Table 1.
List of GM crops grown in different countries with total biotech crop area [5].

of the GM safflower. Since 1996, a total of 2.7 billion hectares of biotech crops are planted. In India, 6 million farmers planted Bt cotton in 11.9 million hectares in 2019. Malawi, Ethiopia, and Nigeria are comparatively new GM crop growers, started from 2019. Interestingly, Japan has approved the cultivation of GM crops, but they have not planted any yet [5]. The top 5 countries to plant biotech crops are USA- 71.5 million hectares (95% adoption), Brazil-52.8 million hectares (94% adoption), Argentina-24 million hectares (~100% adoption), Canada- 12.5 million hectares (90% adoption) and India –11.9 million hectares (94% adoption). According to the report of ISAAA in 2019, biotech soybean is the most cultivated GM crop with 48.2% coverage of the whole biotech crop land, followed by biotech maize (32%), biotech cotton (13.5%), Biotech canola (5.3%) and all other biotech crops cover around 1% of the total biotech crop land [20].

6. Human health risk and safety assessment of genetically modified crops

General people consider foods obtained from conventionally developed crops are safe and GM crops possess risks to human health and the environment. Previously risk assessment used to be carried out based on some common principles that were introduced in 1993 [20]. Later on, an international body called 'Ad Hoc Intergovernmental Task Force on Foods Derived from Biotechnology of the Codex Alimentarius Commission', jointly established by the Food and Agriculture Organization (FAO) and World Health Organization (WHO) and developed a guideline for help the national authorities to assess the human health risk of GM foods. The three main concerns for human health from GM foods are (i) allergenicity—the transgene should not be allergenic (currently no allergic effect was recorded), (ii) gene transfer—from GM food to the body cells or to the bacteria of gut microbiome; particularly concerning if antibiotic resistant genes are used. Therefore, any technique used in GM crop development that involves antibiotic resistant genes are discouraged, and (iii) outcrossing—several cases have been reported where transgene from GM crops approved for animal feed or industrial purposes [21]. According to the Codex Alimentarius, the human health risk assessment of GM crops is based on the principle that a newly developed GM crop should be compared with the conventional counterpart with a history of safe use [22]. The whole idea is that the newly developed GM crop is not compositionally different than its conventional counterpart except the newly introduce gene product/s [23].

According to the studies of National Cancer Institute, from 1975 to 2011, some specific types of cancer incidence has increased and some other has decreased in the USA. No absolute pattern was observed in the incensement or reduction of cancer incidence. If GM foods were a factor to increase cancer incidence in human, then there supposed to be a substantial increase in cancer incidence from 1995 when USA introduced GM crops in the market [24]. Another study published data on cancer incidence worldwide from 1980 to 2010 on breast and cervical cancer. This report showed that the incidence of these 2 cancers has increased in a similar pattern in the USA, Canada, and Europe. Interestingly, GM crops are not generally consumed in the European Union [25]. Epidemiological data was examined in the USA to find out the correlation between GM foods and chronic kidney disease (CKD). The prevalence of all stages CKD in the USA has increased 2% between 1994 and 2004. After that the prevalence did not increase significantly. Therefore, this increase is not correlated with the consumption of GM foods [26]. Similarly, no relationship was found between obesity and type-II diabetes with the consumption of GM foods. Similar conclusion was also made for gastrointestinal tract diseases, celiac disease, and food allergy and autism. Spectrum disorder [27].

The consumption of genetically modified foods direct human health safety assessment is considered as it is safe or unsafe. Many organizations such as the U.S. Environmental Protection Agency (EPA), the U.S. Department of Agriculture (USDA), and the U.S. Food and Drug Administration (FDA) other agencies in other countries or by companies, nongovernmental organizations (NGOs), and academic institutions conducted several research for human health safety assessment. American Association for the Advancement of Science says crop improvement by the modern molecular techniques of biotechnology is safe [28]. Genetically modified foods currently available on the international market have passed safety assessments and are not likely to present risks for human health World Health Organization [21]. All the available GM crops in the market did not show any kind of major human health risks till now. Also, newly developed GM crops must go through proper risk and safety assessments for better acceptability. It suggests the necessity of labeling GM food so that consumers can make their own choice [29].

7. Safety assessment process to register or import GM products

Delivering GM products to market requires time and investment to ensure safety. Use of genetically modified crops as human food and animal feed several studies are required for registration or import into the country. Many countries have very strict rules and regulations for release of GM crops and also a trade barrier in some situations. Hence introduction of GM crops and their products as human food and animal feed some reliable methods such as molecular analysis, safety assessment, toxicological test, nutritional quality analysis, genotoxicity analysis, allergenicity etc. are essential for import or release of new genetically modified crops.

A detailed molecular characterization is a common study in the safety assessment of GM crops.

7.1 Molecular analysis

GM crops can be tested by several molecular levels such as at DNA, mRNA and protein level. Polymerase Chain Reaction (PCR) is the primary method for screening of GM crops at DNA level. This method has found very broad and wide applications in GMO detection and in this method target gene multiplied to millions or billions by using gene specific primers [30]. Southern blotting is another important method for the identification of specific DNA fragments transformed into the genome of transgenic plants or its products which was described by Southern in 1975 [31]. This is very reliable method that provides the molecular evidence of the transgene integration and also estimates the copy number of introduced gene into the host genome. DNA microarray is another method used to identify the expression of more than one gene in a single test. This test method has been used in GMO screening as a method for simultaneous detection of more than 250,000 targets in single assay/chip [32]. In case of RNA based methods real-time PCR, northern blotting techniques etc. are used to monitor and study the gene expression in GM crops. RT-PCR method is based on reverse transcription of mRNA and synthesis of complementary DNA (cDNA) which is then used as template in PCR amplification of target gene. In real-time assay of transgene in GMOs, the amplification and detection occur simultaneously [33]. In northern blotting also requires mRNA as tested material from GMOs. This is a standard method for the analysis of size and level of target RNA in a complex GMO sample. It gives comparative amount of gene expression at the RNA level. Protein based test methods enzyme linked immune sorbent assay (ELISA) and western blot methods have been used for the protein analysis in GMOs. In this assay protein specific antibody coated multi-well plate is used to identify and quantify the specific protein [34].

7.2 Toxicological studies

All toxicity assessment for GM material should be performed based on a case-by-case approach, considering the toxicological profile of new introduced substances [35]. The purpose of toxicological studies is to characterize intended changes and detect active substances or compounds that could have unexpected toxic effects for non-targeted organisms [36]. The assessment of toxicity analysis in animal new strategies have been identify for GM feed and foods. Research on the in-planta metabolism pathway, such as "-omics" techniques that may generate a better understanding of the complex pleotropic effects of new plant cultivars [37]. Additionally, *in vitro* assays with gastric enzymes, cultured cell lines, receptor proteins, and *in vivo* animal studies can be performed [36]. When performing *in vivo* studies, toxicology acute (14 days studies), subacute (28 days studies), chronic (90 days studies), or specific toxicity (reproductive, mutagenicity, etc.) assessment can be considered [38]. In long term studies (>100 days), no toxic effects were found in cattle and chickens fed with Bt maize. The toxicological evaluation of proteins introduced into GM crops must be carried out for introduced or release of GMO crops and carefully assess the modification of amino acid sequences that make a non-toxic protein to toxic protein [39].

7.3 Nutritional studies of GMO crops

It is important to assess the nutritional quality of genetically modified foods that consumed in human and animal feed. Assessment of nutritional quality composition of transgenic wheat (*Triticum durum* L.), corn (*Zea mays* L.), and tomato (*Lycopersicum esculentum* Mill.) were done compare with the non-transgenic control with a similar genetic background. No significant differences were observed for qualitative traits analyzed in wheat and corn samples [40]. Many studies have also been carried out with feed derived from GM plants with agronomic input traits in target animal species to assess the nutritive value of the feed and their performance potential. Studies in sheep, pigs, broilers, lactating dairy cows, and fish, comparing the *in vivo* bioavailability of nutrients from a range of GM plants with their near isogenic counterpart and commercial varieties, showed that they were comparable with those for near isogenic non-GM lines and commercial varieties [41]. In case of proteomic analysis creation of respective databases and algorithms of plant proteomic analysis is among the most important avenues of fundamental science aimed to predict the functions of genes and the properties of the products encoded by them [42].

7.4 Assessment of potential genotoxicity

Assessment of potential genotoxicity of the foods obtained from GM crops for human health safety assessment is important. Genotoxic studies include the assessment of genetic material at various stages of DNA, chromosomes etc. [43]. Accordance with the "Food Additive Risk Assessment Guidelines" of the Japan Food Safety Commission were assessed the genotoxicity of 30 food-flavoring chemicals in GM crops. Of the 30 food-flavoring chemicals, three yielded a positive result in both Ames and CA tests [44]. Another 11 chemicals yielded positive results in the CA test. However, none of the chemicals yielding positive *in vitro* test results yielded positive results in the *in vivo* tests. These findings indicate no genotoxicity concerns of the food-flavoring chemicals belonging to the above mentioned 18 chemical classes used in Japan unless there are other structural modifications.

7.5 Assessment of potential allergenicity

The assessment of allergenicity reaction is important to release a GMO crop. A variety of factors considered for an overall assessment of allergenic potential is conducted. This assessment includes a safety of the genes source, protein structure (e.g. amino acid sequence identity to human allergens) and stability of the protein to pepsin digestion *in vitro*, heat stability of the protein, glycosylation status, and specific IgE binding studies with sera from relevant clinically allergic subjects [45]. Since GM crops were first commercialized over 20 years ago, there is no proof that the introduced novel protein(s) in any commercialized GM food crop has caused food allergy. The process by which allergy assessment has been conducted since the 1990s has involved guidance from several expert scientific bodies, including the FAO/WHO [46], and Codex [22]. However, several animal models are important for investigating the etiology of food allergy as well as evaluating advances in immunotherapy techniques to induce desensitization and ultimately, tolerance to food allergic reactions [47]. Now days, Bioinformatics tools has led to advances in predicting the allergenicity of novel proteins.

7.6 Analysis of sensitive biomarkers

Some marker genes are introduced into the gene construct for identification of transgenic plants. So, it is necessary to assess the effect of biomarker in human or other plants. A machine learning methodology would enable the identification of potential biomarkers associated with the potential adverse health effects related to GMO exposure. This broad approach will allow for the collection and analysis of vast quantities of data for biomarker identification. Researchers were not successful in their search for biomarkers cause any health effects [48].

8. Methods of identification and detection of GMO

The number and diversity of genetically modified organisms (GMOs) for the food and feed market is increasing day by day. So, it is important to efficient identification and detection of GMO in food and feed products. Several strategies have been developed to detect GMO in food/feed samples by using different technologies. In case of the protein-based approaches, ELISA technique and mass spectrometry-based technology are used to characterize GM crops [49, 50]. Protein based detection method is not applicable if the genetic modification has no impact at the protein level. Currently, Many DNA based GMO detection methods have been developed such as quantitative PCR (qPCR), loopmediated isothermal amplification (LAMP), PCR capillary gel electrophoresis (CGE), microarray, Luminex, digital PCR (dPCR), DNA walking and Next Generation Sequencing (NGS).

8.1 qPCR technology

Event-specific PCR detection technology is commonly employed for GMO testing due to its ability to specifically detect each transgenic event simply by targeting their unique junction between the host genome and the transgenic cassette [51]. Currently, different event-specific qPCR (quantitative) technology has been designed for transgene detection from GM Corn, Cotton, Canola, Rapeseed, and rest of the crops [52]. The qPCR system, which is the most common strategy, allows detecting, identifying, and quantifying GMO via the SYBR Green or TaqMan chemistries in agricultural and food products [53]. Though, the (qPCR) methods have more reliable, accuracy and greater sensitivity but its success largely relies on various factors, e.g., its throughput strategy is often restricted to one marker per reaction. Due to continuous growth in GMO production, new/additional detection markers (for specific detection of new transgene) are required to be designed continuously and used to completely cover their identification. In case of multiplex PCR-based methods, several DNA targets can be detected in a single reaction.

8.2 Capillary gel electrophoresis (CGE)

CGE technique was developed by Heide et al. to identify various transgenic events in one reaction [54]. The basic principle of this technique is to carry out multiple PCR reaction using forward primers which are fluorescently labeled and discrimination of amplimer of similar magnitude by executing CGE. As compared to the electrophoresis gel, CGE system has higher resolution power to clearly detect PCR products from a multiplex assay. Using the CGE system, eight GM maize were identified via a nonaplex PCR including event-specific, construct-specific, and taxon-specific methods

[55]. Similarly, one pentaplex PCR and two hexaplex PCR were also developed to detect specifically four GM maize and five GM cotton [56]. However, CGE has some disadvantages as it requires extensive labor for designing of primer as well as the optimization when performing the analyses for detection of a new event. Its implementation also requires specialized apparatus which may not always be available.

8.3 Loop mediated isothermal amplification (LAMP)

Loop mediated isothermal amplification (LAMP) is an emerging technology which was developed an easy detection of the transgenic event in a given sample [57]. The main idea is the amplification and identification of the desired nucleic acid sequences at a steady temperature and at some specific stage of the experiment. This novel approach of GMO detection involves the utilization of four distinct primers which identify at least six different segments of the desired DNA. Varieties of LAMP markers were designed for quantitative detection of transgenic GM events [58]. The advantage of this study is simplicity, time-efficiency, and ability to withstand different PCR inhibitors, for instance, acidic polysaccharides. However, this technique has some limitations, e.g., designing four primers per sequence and detection of different GM events employing multiplex approach is also a problem.

8.4 Digital PCR (dPCR)

Digital PCR technology (dPCR) is one of the most reliable techniques among the currently used technology for GMO quantification. The process is accomplished through dividing the mixture of PCR into a sizeable number of distinct reactions which include null, single or least target DNA copies. After completion of PCR, the positive (i.e., observed replicated desired segments) and negative (i.e., observed non-replicated segments) samples are analyzed and then the total copy number of the desired gene in an original sample is determined by the application of binomial Poisson statistics [59]. Most recently, duplex assays, including one GMO specific marker with one soybean, maize, or rice taxon specific marker, were performed by using the dPCR system to quantify 12 GM soybean, 16 GM maize, and two GM rice events [60].

8.5 Microarray technology

A microarrays technique (DNA chips or biochips) is an advanced technology for high-throughput detection of GMO. It can evolve together with the growing number of newly developed GMO in the food and feed markets. With the microarray technology applied to GMO detection, GM targets are amplified by PCR, using target-specific and/or universal primers, prior to being hybridized on the array, allowing the simultaneous detection of more than 250,000 targets in one assay [61]. DNA chip technology coupled with multiplex PCR can be used in the identification of different transgenic events from GMOs by employing multiplex PCR approaches. Nucleic acid array in combination with multiplex PCR has been used successfully for identification of different types of events from GM crops like corn and cotton [32].

8.6 Next generation sequencing (NGS)

Next generation sequencing (NGS) is a promising technology with the detection of transgenic events that allows for massively parallel DNA segment sequencing resulting in millions of sequencing reads [62]. NGS is an efficient tool for transgenic events detection even in the absence of sequence information of such events. NGS is being efficiently employed for characterization of site addition, flanking regions, accidental addition as well as the determination of transgene copy number [63]. Two main approaches are (targeted sequencing strategy) or (whole genome sequencing (WGS) strategy), for samples sequencing which has been enriched previously with desire sequence regions have been identified [64]. Molecular analyses of transgenic varieties from GM soy and GM rice have successfully been achieved using this strategy [65]. An approach based on initial enrichment of the targets using DNA walking starting from transgenic elements covering a large spectrum of GMO has been proposed to characterize unknown sequences including the transgenic inserts and junctions [66]. DNA walking technology will become more sensitive and more suitable and that could provide a more promising solution for the recent challenges of GMO analysis in the near future.

9. GM crops for food security

With the current population growth, it is expected that the world population will reach 9 billion by 2050 and current food production rate is not sufficient to feed the future population [67]. On top of that, 40% of the total arable land of the world is lost over the last 40 years because of erosion and pollution. Heavy utilization of arable lands reduces organic matter and nutrients from the soil; therefore, farmers need to rely on heavy fertilizer input to reintroduce nutrients in the soil for better crop yield [68]. In addition to heavy input of synthetic fertilizers, farmers also need to rely on insecticides and pesticides as every year farmers face 20–30% due to insects, animals, and weeds [69, 70]. Therefore, the most sustainable way to ensure food security is to develop high yielding, enhanced product quality, disease resistance and climate resilient crop varieties. Plant breeders try to develop new superior plant varieties; however, it is limited by the huge number of crosses needed to be done, prolonged and labor intense selection process, transfer of undesired genes with the desired ones, random combination of genes from two parents and so on [71]. In contrast, GM crops are developed by incorporating only the desired gene/s in the target crop, even from a distantly related organism, in a shorter period of time and in more precise manner. GM crops have a lot more to offer than the conventionally developed varieties. Biotech crops are already proven to be more economically beneficial as it requires less chemical input such as herbicide tolerant and insect tolerant GM crops. GM crops are the fastest adopting technology in the world. If farmers have not decided to cultivate GM crops, in order to obtain present day crop yield, we would have needed an additional 13 million hectares of land [72]. This amount of land would come at the cost of destroying rainforests which would eventually increase carbon emission. Most of the biotech crops are cost saving and some are with higher yield capacity. As a result, the farmers are getting more return from the GM crops [73]. GM crops have all the potential to achieve food security, however, lack of societal acceptance of GM crops is the major obstacle as many farmers fear that if they grow GM crops then consumers may not buy it. Also, most of the people mistrusts the benefits of GM crops because they think GM crops are actually harmful, but developers of new GM technologies and GM varieties hides for profit [68]. In real life, this is not the case. Thousands of studies are present that showed GM crops are not harmful at all. It is the duty of the scientific community to make general people understand the science and safety of the GM crops in an easier and clearer manner.

10. Bio-safety assessments of genetically modified food

Biosafety refers to the safe management of living organisms and genetic material, including pathogens and genetically modified organisms. Under international environmental law and policy, biosafety refers to the need to protect the environment and human health from the possible adverse effects of genetically modified organisms (GMOs) and products resulting from modern biotechnology. Currently, at the core of the international regime on biosafety is the 2000 Cartagena Protocol on Biosafety, adopted under the Convention on Biological Diversity (CBD). Now a days dozens of genetically modified (GM) food crops and animals have been developed and commercialized in different countries. That's why several biosafety concerns like- risk to human health, risk to environment, ecological concern has been raised after the rapid commercialization of GM crops every year across the world. So strict biosafety guideline must be followed before introducing a GM crop into the market. According to Cartagena Protocol on Biosafety (CPB) take appropriate measures to regulate, manage or control the risks that may arise due to use and handling of living modified organisms (LMOs) that may pose some threats to biological and to ensure the safe handling, transport, and use of LMOs. Different countries and organizations developed their own biosafety guidelines for safety assessment of genetically modified foods. The safety assessment is undertaken in accordance with internationally established scientific principles and guidelines developed through the work of the Organization for Economic Cooperation and Development (OECD), Food and Agriculture Organization (FAO) of the United Nations, World Health Organization (WHO) and the Codex Alimentarius Commission. Risk assessment must be completed on the basis of scientific evidence to identify and evaluate the possible impacts of GMO on the conservation and sustainable use of biodiversity as well as risks to human health [74]. Risk assessment identifies potential hazards and/or adverse impacts of GM crops or derived product on non-target organisms and/or environment. This involves a number of coordinated steps like risk identification, risk characterization and risk categorization. Risk management involves strategic techniques to reduce the adverse effect of GM crops and associated products on non-target species or environment and also to reduce the chances of development of resistance in target pest population [75]. Risk management is a follow-up to the implementation of a risk assessment that includes the establishment of appropriate mechanisms, steps, and strategies for managing and controlling the risks identified in the risk assessment. The obligation arising from the application of risk management to these parties is to establish and implement a regulatory system with sufficient capacity to manage and control these risks. Regulation of GMO handling, transport, packaging, and utilization is part of efforts to ensure the safety of GMO development in accordance with the requirements of relevant international standards. In case of USA, the U.S. Food and Drug Administration (FDA), U.S. Environmental Protection Agency (EPA), and U.S. Department of Agriculture (USDA) ensure that GMOs are safe for human, plant, and animal health [76]. Safety assessment of GM foods should be carried out on a case-by-case basis comparing the properties of new food with those of conventional counterpart.

11. Conclusion

Genetic modified crops have all the potential for ensuring food security. There is a misconception present surrounding the GM crops that it is harmful for the nature and

GM food brings harms to human health including cancer. This is due to total ignorance towards the scientific proofs that GM crops do not possess any of these threats under proper guidelines. The organic foods do not require synthetic fertilizer or insecticides or herbicides, but it is limited by the low productivity. It is easily understandable that organic farming cannot meet the required crops yield. High yielding varieties developed through conventional breeding methods, most of the time require heavy synthetic chemical inputs, including herbicides, pesticides, fertilizers. The harms that are brought to the environment through the heavy application of such synthetic chemicals is far more harmful. In recent years, GM crops with stacked traits has become more popular. In conventional ways, it requires a lot of time to develop a variety with multiple desired characteristics. That means GM crops can prepare us for future food security challenges. The GM crops need to be gone through proper regulations before being introduced in the market including health and environmental risk assessments. By far no GM crops available in the market for commercial use showed any signs of harm to human health. GM foods coming from GM crops are also properly regulated and labeled. The scientific community must work hard to make general people understand about this technology and its benefits to eradicate the misconceptions about GM crops and foods, whereas regulatory bodies should work through proper guidelines to gain people's trust so they can be attracted to GM crops or GM foods. Therefore, it can be concluded that with proper social acceptance, genetic modified crops can ensure food security for the future world.

Author details

Sujay Kumar Bhajan[1*†], Mohammad Mahmood Hasan[2†], Md. Anwarul Haque[3] and Mohammad Nurul Islam[4]

1 Department of Microbiology, Rabindra Maitree University, Kushtia, Bangladesh

2 Department of Microbiology and Cell Science, University of Florida, Gainesville, FL, USA

3 Department of Biotechnology and Genetic Engineering, Islamic University, Kushtia, Bangladesh

4 Plant Breeding and Biotechnology Laboratory, Department of Botany, University of Dhaka, Dhaka, Bangladesh

*Address all correspondence to: skbhajan.tech@gmail.com

† These authors contributed equally to this work.

References

[1] World Hunger. 2013. Available from: https://www.worldhunger.org/articles/Learn/old/world%20hunger%20facts%202002_2012version.htm

[2] Hanjra MA, Qureshi ME. Global water crisis and future food security in an era of climate change. Food Policy. 2010;**35**:365-377

[3] Barros J, Temple S, Dixon RA. Development and commercialization of reduced lignin alfalfa. Current Opinion in Biotechnology. 2019;**56**:48-54. DOI: 10.1016/j.copbio.2018.09.003

[4] Halford NG, Shewry PR. Genetically modified crops: Methodology, benefits, regulation and public concerns. British Medical Bulletin. 2000;**56**(1):62-73. DOI: 10.1258/0007142001902978

[5] International Service for the Acquisition of Agri-biotech Applications (ISAAA). Global Status of Commercialized Biotech/GM Crops in 2019: Biotech Crops Drive Socio-Economic Development and Sustainable Environment in the New Frontier. Ithaca, New York: ISAAA Brief; 2019

[6] Brookes G, Barfoot P. GM Crops: The First Ten Years-global Socioeconomic and Environmental Impacts. Dorchester, United Kingdom: PG Economics Limited; 2006

[7] van den Eede G, Aarts H, Buhk HJ, Corthier G, Flint HJ, Hammes W, et al. The relevance of gene transfer to the safety of food and feed derived from genetically modified (GM) plants. Food and Chemical Toxicology. 2004;**42**(7):1127-1156. DOI: 10.1016/j.fct.2004.02.001

[8] Low L-Y, Yang S-K, Kok D, Ong-Abdullah J, Tan NP, Lai K-SJNViPS. Transgenic plants: Gene constructs, vector and transformation method. 2018

[9] Singh T, Singh T, Singh R, Gaur R, Pandey K, Jamal F. Genetic engineering: Altering the threads of life. 2020. pp. 1-11

[10] Pratiwi R, Surya MI. Agrobacterium-mediated transformation. In: To K-Y, editor. Genetic Transformation in Crops. London: IntechOpen; 2020

[11] Sanford JC. Biolistic plant transformation. 1990;**79**(1):206-209. DOI: 10.1111/j.1399-3054.1990.tb05888.x

[12] Naylor LH. Reporter gene technology: The future looks bright. Biochemical Pharmacology. 1999;**58**(5):749-757. DOI: 10.1016/s0006-2952(99)00096-9

[13] Dai S, Zheng P, Marmey P, Zhang S, Tian W, Chen S, et al. Comparative analysis of transgenic rice plants obtained by Agrobacterium-mediated transformation and particle bombardment. Molecular Breeding. 2001;**7**(1):25-33. DOI: 10.1023/A:1009687511633

[14] Bubner B, Baldwin IT. Use of real-time PCR for determining copy number and zygosity in transgenic plants. Plant Cell Reports. 2004;**23**(5):263-271. DOI: 10.1007/s00299-004-0859-y

[15] Lambirth KC, Whaley AM, Schlueter JA, Bost KL, Piller KJ. CONTRAILS: A tool for rapid identification of transgene integration sites in complex, repetitive genomes using low-coverage paired-end sequencing. Genom Data. 2015;**6**:175-181. DOI: 10.1016/j.gdata.2015.09.001

[16] Food and Drug Administration (FDA). Science and History of GMOs and Other Food Modification Processes. 2022. Available from: https://www.fda.gov/food/agricultural-biotechnology/science-and-history-gmos-and-other-food-modification-processes. [Accessed: June 25, 2022]

[17] International Service for the Acquisition of Agri-biotech Applications (ISAAA), GM Approval Database. 2022. Available from: https://www.isaaa.org/gmapprovaldatabase/. [Accessed: June 25, 2022]

[18] Brijesh P, Rajesh K, Abhishek KSJSR. Essays, expanding the global horizon of engineered crops: Miles ahead to go. 2018

[19] Aldemita RR, Reaño IME, Solis RO, Hautea RA. Trends in global approvals of biotech crops (1992-2014). GM Crops & Food. 2015;**6**(3):150-166. DOI: 10.1080/21645698.2015.1056972

[20] Organisation for Economic Co-operation and Development (OECD). Safety Evaluation of Foods Derived by Modern Biotechnology: Concepts and Principles. Paris, France: Organization for Economic Co-operation and Development; 1993

[21] World Health Organization (WHO). Frequently Asked Questions; Food, Genetically Modified. 2014. Available from: https://www.who.int/news-room/questions-and-answers/item/food-genetically-modified. [Accessed: June 25, 2022]

[22] Food and Agriculture Organization (FAO) and World Health Organization (WHO). Codex Alimentarius. 2003. Available from: https://www.fao.org/fao-who-codexalimentarius/thematic-areas/biotechnology/en/. [Accessed: June 25, 2022]

[23] Schauzu M. The concept of substantial equivalence in safety assessment of foods derived from genetically modified organisms. AgBiotechNet. 2000;**2**(044):1-4

[24] National Cancer Institute (NCI). Cancer Facts & Figures 2015. 2015. Available from: https://www.cancer.org/research/cancer-facts-statistics/all-cancer-facts-figures/cancer-facts-figures-2015.html. [Accessed: June 25, 2022]

[25] Forouzanfar MH, Foreman KJ, Delossantos AM, Lozano R, Lopez AD, Murray CJ, et al. Breast and cervical cancer in 187 countries between 1980 and 2010: A systematic analysis. Lancet. 2011;**378**(9801):1461-1484. DOI: 10.1016/s0140-6736(11)61351-2

[26] The United States Renal Data System (USRDS). CKD in the general population. USRDS Annual Data Report Volume, USA. 2014:21-22

[27] National Academies of Sciences. Engineering, and Medicine, Human Health Effects of Genetically Engineered Crops, Genetically Engineered Crops: Experiences and Prospects. Washington D.C., USA: National Academies Press; 2016

[28] American Association for Advancement of Sciences (AAAS). Statement by the AAAS Board of Directors on Labeling of Genetically Modified Foods. 2013. Available from: https://www.aaas.org/news/statement-aaas-board-directors-labeling-genetically-modified-foods. [Accessed: June 28, 2022]

[29] Shen C, Yin X-C, Jiao B-Y, Li J, Jia P, Zhang X-W, et al. Evaluation of adverse effects/events of genetically modified food consumption: A systematic review of animal and human studies. Environmental Sciences

Europe. 2022;**34**(1):8. DOI: 10.1186/s12302-021-00578-9

[30] Mullis KB, Faloona FA. Specific synthesis of DNA in vitro via a polymerase-catalyzed chain reaction. Methods in Enzymology. 1987;**155**:335-350. DOI: 10.1016/0076-6879(87)55023-6

[31] Southern EM. Detection of specific sequences among DNA fragments separated by gel electrophoresis. Journal of Molecular Biology. 1975;**98**(3):503-517. DOI: 10.1016/s0022-2836(75)80083-0

[32] Kim J-H, Kim S-Y, Lee H, Kim Y-R, Kim H-Y. An event-specific DNA microarray to identify genetically modified organisms in processed foods. Journal of Agricultural and Food Chemistry. 2010;**58**(10):6018-6026. DOI: 10.1021/jf100351x

[33] Higuchi R, Fockler C, Dollinger G, Watson R. Kinetic PCR analysis: Real-time monitoring of DNA amplification reactions. Biotechnology (N Y). 1993;**11**(9):1026-1030. DOI: 10.1038/nbt0993-1026

[34] Kuiper H. Summary report of the ILSI Europe workshop on detection methods for novel foods derived from genetically modified organisms. Food Control. 1999;**10**:339-349

[35] Domingo JL. Toxicity studies of genetically modified plants: A review of the published literature. Critical Reviews in Food Science and Nutrition. 2007;**47**(8):721-733. DOI: 10.1080/10408390601177670

[36] Van Haver E, De Schrijver A, Devos Y, Lievens S, Renckens S, Moens WJ. The safety assessment of genetically modified crops for food and feed use. In: Guidance Notes fromthe Service of Biosafety Biotechnology Biosafety Council in Belgium. 2003 Report of the Belgium Biosafety Advisory Council, Scientific Institute of Public Health. Royal Library of Belgium Deposit No. D/2003/2505/16., Belgium

[37] Fernandez A, Paoletti C. Unintended effects in genetically modified food/feed safety: A way forward. Trends in Biotechnology. 2018;**36**(9):872-875. DOI: 10.1016/j.tibtech.2018.03.005

[38] Levitsky E. Problem of genetically modified foods safety: A toxicologist's view. Biotechnologia Acta. 2016;**9**:7-25. DOI: 10.15407/biotech9.01.007

[39] Hammond B, Kough J, Herouet-Guicheney C, Jez JM. Toxicological evaluation of proteins introduced into food crops. Critical Reviews in Toxicology. 2013;**43**(Suppl. 2):25-42. DOI: 10.3109/10408444.2013.842956

[40] Venneria E, Fanasca S, Monastra G, Finotti E, Ambra R, Azzini E, et al. Assessment of the nutritional values of genetically modified wheat, corn, and tomato crops. Journal of Agricultural and Food Chemistry. 2008;**56**(19):9206-9214. DOI: 10.1021/jf8010992

[41] EFSA GMO Panel Working Group on Animal Feeding Trials. Safety and nutritional assessment of GM plants and derived food and feed: The role of animal feeding trials. Food and Chemical Toxicology. 2008;**46**(Suppl. 1):S2-S70. DOI: 10.1016/j.fct.2008.02.008

[42] König A, Cockburn A, Crevel RW, Debruyne E, Grafstroem R, Hammerling U, et al. Assessment of the safety of foods derived from genetically modified (GM) crops. Food and Chemical Toxicology. 2004;**42**(7):1047-1088. DOI: 10.1016/j.fct.2004.02.019

[43] Tutel'jan VA, Gapparov MG, Avren'eva LI, Aksjuk IN, Guseva GV. Medico-biological assessment of the

safety of genetically engineered maize line MON 880 Communication Toxicological and hygienic studies. Problems of Nutrition. 2008;**5**:4-13

[44] Honma M, Yamada M, Yasui M, Horibata K, Sugiyama K-I, Masumura K. Genotoxicity assessment of food-flavoring chemicals used in Japan. Toxicology Reports. 2022;**9**:1008-1012. DOI: 10.1016/j.toxrep.2022.04.026

[45] Ladics GS. Assessment of the potential allergenicity of genetically-engineered food crops. Journal of Immunotoxicology. 2019;**16**(1):43-53. DOI: 10.1080/1547691X.2018.1533904

[46] World Health Organization (WHO). Safety aspects of genetically modified foods of plant origin: Report of a joint FAO/WHO expert consultation on foods derived from biotechnology. 2000

[47] Larsen JM, Bøgh KL. Animal models of allergen-specific immunotherapy in food allergy: Overview and opportunities. Clinical and Experimental Allergy. 2018;**48**(10):1255-1274. DOI: 10.1111/cea.13212

[48] Aksyuk IN, Chernysheva ON, Kravchenko LV, Mazo VK, Onishchenko GG, Rogov IA, et al. Principles of human health safety assessment of genetically modified plants used in the Russian Federation. In: Tutelyan VA, editor. Genetically Modified Food Sources. San Diego: Academic Press; 2013. pp. 31-42

[49] García-Cañas V, Simó C, León C, Ibáñez E, Cifuentes A. MS-based analytical methodologies to characterize genetically modified crops. Mass Spectrometry Reviews. 2011;**30**(3):396-416. DOI: 10.1002/mas.20286

[50] Lobato IM, O'Sullivan CK. Recombinase polymerase amplification: Basics, applications and recent advances. Trends in Analytical Chemistry. 2018;**98**:19-35. DOI: 10.1016/j.trac.2017.10.015

[51] Zhang M, Yu Y, Gao X, Zhang K, Luan F, Zhu Y, et al. Event-specific quantitative detection of genetically modified wheat B72-8-11 based on the 3′ flanking sequence. European Food Research and Technology. 2015;**240**(4):775-782. DOI: 10.1007/s00217-014-2383-9

[52] Jiang L, Yang L, Rao J, Guo J, Wang S, Liu J, et al. Development and in-house validation of the event-specific qualitative and quantitative PCR detection methods for genetically modified cotton. MON15985. 2010;**90**(3):402-408. DOI: 10.1002/jsfa.3829

[53] Angers-Loustau A, Petrillo M, Bonfini L, Gatto F, Rosa S, Patak A, et al. JRC GMO-matrix: A web application to support Genetically Modified Organisms detection strategies. BMC Bioinformatics. 2014;**15**(1):417. DOI: 10.1186/s12859-014-0417-8

[54] Heide BR, Heir E, Holck A. Detection of eight GMO maize events by qualitative, multiplex PCR and fluorescence capillary gel electrophoresis. European Food Research and Technology. 2008;**227**(2):527-535. DOI: 10.1007/s00217-007-0751-4

[55] Heide BR, Drømtorp SM, Rudi K, Heir E, Holck AL. Determination of eight genetically modified maize events by quantitative, multiplex PCR and fluorescence capillary gel electrophoresis. European Food Research and Technology. 2008;**227**(4):1125-1137. DOI: 10.1007/s00217-008-0828-8

[56] Holck A, Pedersen BO, Heir E. Detection of five novel GMO maize

events by qualitative, multiplex PCR and fluorescence capillary gel electrophoresis. European Food Research and Technology. 2010;**231**(3):475-483. DOI: 10.1007/s00217-010-1302-y

[57] Li Q, Fang J, Liu X, Xi X, Li M, Gong Y, et al. Loop-mediated isothermal amplification (LAMP) method for rapid detection of cry1Ab gene in transgenic rice (*Oryza sativa* L.). European Food Research and Technology. 2013;**236**(4):589-598. DOI: 10.1007/s00217-013-1911-3

[58] Zahradnik C, Kolm C, Martzy R, Mach RL, Krska R, Farnleitner AH, et al. Detection of the 35S promoter in transgenic maize via various isothermal amplification techniques: A practical approach. Analytical and Bioanalytical Chemistry. 2014;**406**(27):6835-6842. DOI: 10.1007/s00216-014-7889-2

[59] Fraiture M-A, Herman P, Taverniers I, De Loose M, Deforce D, Roosens NH. Current and new approaches in GMO detection: Challenges and solutions. BioMed Research International. 2015;**2015**:392872. DOI: 10.1155/2015/392872

[60] Paternò A, Verginelli D, Bonini P, Misto M, Quarchioni C, Dainese E, et al. In-house validation and comparison of two wheat (*Triticum aestivum*) taxon-specific real-time PCR methods for GMO quantification supported by droplet digital PCR. Food Analytical Methods. 2018;**11**(5):1281-1290. DOI: 10.1007/s12161-017-1097-6

[61] Nakaya HI, Reis EM, Verjovski-Almeida S. Concepts on microarray design for genome and transcriptome analyses. In: Buzdin AA, Lukyanov SA, editors. Nucleic Acids Hybridization Modern Applications. Netherlands, Dordrecht: Springer; 2007. pp. 265-307

[62] Willems S, Fraiture M-A, De Keersmaecker S, Roosens NH. Next generation sequencing to identify GMO in food and feed products. Labinfo. 2015;**13**:18-20

[63] Milavec M, Dobnik D, Yang L, Zhang D, Gruden K, Žel J. GMO quantification: Valuable experience and insights for the future. Analytical and Bioanalytical Chemistry. 2014;**406**(26):6485-6497. DOI: 10.1007/s00216-014-8077-0

[64] Fraiture M-A, Herman P, De Loose M, Debode F, Roosens NH. How can we better detect unauthorized GMOs in food and feed chains? Trends in Biotechnology. 2017;**35**(6):508-517. DOI: 10.1016/j.tibtech.2017.03.002

[65] Yang L, Wang C, Holst-Jensen A, Morisset D, Lin Y, Zhang D. Characterization of GM events by insert knowledge adapted re-sequencing approaches. Scientific Reports. 2013;**3**(1):2839. DOI: 10.1038/srep02839

[66] Fraiture M-A, Saltykova A, Hoffman S, Winand R, Deforce D, Vanneste K, et al. Nanopore sequencing technology: A new route for the fast detection of unauthorized GMO. Scientific Reports. 2018;**8**(1):7903. DOI: 10.1038/s41598-018-26259-x

[67] Fedoroff NV, Battisti DS, Beachy RN, Cooper PJM, Fischhoff DA, Hodges CN, et al. Radically rethinking agriculture for the 21st century. Science. 2010;**327**(5967):833-834. DOI: 10.1126/science.1186834

[68] Tyczewska A, Woźniak E, Gracz J, Kuczyński J, Twardowski TJ. Towards food security: Current state and future prospects of agrobiotechnology. Trends in Biotechnolgy. 2018;**36**(12):1219-1229

[69] Culliney T. Crop losses to arthropods. In: Pimentel D, Peshin R, editors.

Integrated Pest Management. Dordrecht: Springer; 2014. DOI: 10.1007/978-94-007-7796-5_8

[70] Pandey P, Irulappan V, Bagavathiannan MV, Senthil-Kumar M. Impact of combined abiotic and biotic stresses on plant growth and avenues for crop improvement by exploiting physio-morphological. Traits. 2017:8. DOI: 10.3389/fpls.2017.00537

[71] Shukla M, Al-Busaidi KT, Trivedi M, Tiwari RK. Status of research, regulations and challenges for genetically modified crops in India. GM Crops & Food. 2018;**9**(4):173-188. DOI: 10.1080/21645698.2018.1529518

[72] Barrows G, Sexton S, Zilberman D. The impact of agricultural biotechnology on supply and land-use. Environment Development Economics. 2014;**19**(6):676-703

[73] Brookes G, Barfoot P. GM crop technology use 1996-2018: Farm income and production impacts. GM Crops & Food. 2020;**11**(4):242-261. DOI: 10.1080/21645698.2020.1779574

[74] Eggers B, Mackenzie R. The cartagena protocol on biosafety. Journal of International Economic Law. 2000;**3**(3):525-543

[75] Kumar A, Jaiswal P, Janeja HS. Biosafety aspects of genetically modified crops. Genetically Modified Plants and Beyond. 2022;(1-5) IntechOpen. https://doi.org/10.5772/int echopen.1017

[76] Food and Drug Administration (FDA). How GMOs Are Regulated for Food and Plant Safety in the United States. 2022. Available from: https://www.fda.gov/food/agricultural-biotechnology/how-gmos-are-regulated-food-and-plant-safety-united-states [Accessed: June 28, 2022]

Section 3

Agricultural Practices for Food Security

Chapter 5

Phosphate Solubilization Improvement for Plant Uptake from Phosphate Rock and Phosphate Solubilizing Microbes Consortium: Impact on Food Security

Zainab Muhammad Bello, Sanusi Muhammad, Adamu Aliyu Aliero, Adamou Rabani and Ibrahim Aliyu Dabai

Abstract

Poverty alleviation in Sub-Saharan Africa is proportionally dependent on soil management. Low crop productivity has been linked to hunger and poverty as soil degradation is undeniably the cause. This chapter gives a general overview from major findings on how microbes could improve phosphate (P) levels in soils by enhancing its solubility. A cross-sectional study was under taken to highlight the role played by phosphate-solubilizing microbes—arbuscular mycorhizal fungi (AMF) and phosphate-solubilizing bacteria (PSB) in improving phosphate solubility. About 30–50% of phosphorus is organic which the plants could readily assimilate, while 50–70% is inorganic and inaccessible to plants. There are several mechanisms the plants utilize to optimize nutrient uptake from the root hairs to various parts of the plant to maximize crop production. The utilization of readily available minerals such as phosphate rock is known to play vital role in plant ecology and evolution, in checking drought stress, heavy metal toxicity, nutritional imbalances, plant pathogens, and salinity. Therefore, soil improvement using rock phosphate could potentially act in synergy with the phosphate-solubilizing microbes to boost phosphate levels in the soil. This could be a welcome development in low-income economies in the sub-Saharan Africa (SSA) to boost yield for profit maximization.

Keywords: phosphate rock, phosphate-solubilizing microbes, sub-Saharan Africa, food security, plant uptake

1. Introduction

Food production is crucial for our existence and every other life on earth. The rate at which phosphorus (P), a critical ingredient in growing food, is declining is alarming. It was estimated that existing rock phosphate reserves could be exhausted in the next 50–100 years [1–3]. Phosphorus is the second most likely nutrient deficient in the soils after nitrogen [3]. It is one of the most important nutrients for crop production. P availability is usually low in soils around the globe. Moreover, the low efficacy of P fertilizers in acidic and calcareous soils restricts P availability [4]. Recently, it has become increasingly difficult for small-scale farmers in developing countries to purchase chemical P fertilizers.

Rock phosphate is a sedimentary rock that contains high amounts of phosphorus. The rock is mined and in it contains clay and limestone [5]. Rock phosphate had a long history and had been used as organic fertilizer for gardens. It's known for keeping plants healthy and encouraging new growth [5]. Rock phosphate can be used as crude phosphatic fertilizer by direct application to field soil. It was established that double application of RP and phosphate-solubilizing fungi/bacteria improve P content in soil [6]. Moreover, the farmers in sub-Saharan Africa (SSA) are facing high prices of phosphate fertilizers because of the low solubility of the African RPs. Therefore, an affordable P fertilizer supply from local low-grade phosphate rocks would strengthen crop production in SSA and improve farmers' income.

Fungi play fundamental roles in regulating key ecosystem processes such as decomposition of organic matter and plant–soil relationship [7]. Arbuscular mycorrhizal fungi (AMF) on the other hand are widespread obligate plant symbionts that can colonize the roots of most land plants. They also assist in obtaining nutrients and protection against environmental stresses [8]. It was similarly reported that AMF symbiosis improves plant stress resistance and soil stability, making it a promising addition to sustainable agricultural practices [9]. It's crucial to develop a method to enhance P solubility in African degraded soils. There have been significant positive reports on application of partially acidulated RPs on crop cultivation including pearl millet, sorghum, cowpea and maize. It's well known that calcination of RP with Na carbonate increases its solubility, but this method is a classical option, which has been employed for ages in solubilizing low-graded RPs. It's well known that calcination with Na carbonate increases its solubility but proven difficult due to high content of impurities such as silicates [10].

There is need to develop knowledge on natural biological agents that may sustainably improve crop performance and lessen our reliance on technology [11]. AMF symbiosis may be optimized to improve the sustainability of agricultural systems by increasing the ability of crops to absorb soil P, resist pathogens, and tolerate drought stress [12]. AMF are among the microbial groups that could solubilize mineral phosphates and improve plant phosphorus nutrition. AMF inoculation was also reported to induce spectacular stimulations of the plant growth and phosphorus foliar content [13].

2. Contributions of phosphate-solubilizing bacteria (PSB) and arbuscular mycorrhizal fungi (AMF) in improving P availability

A better understanding of how the beneficial microbes (PSB and AMF) and phosphate rock interact is crucial to preserving the soil fertility and improving the economic and environmental sustainability of crop production in P-deficient soils [14]. Previous studies have reported that SiO_2 is not the major constituent of RP but CaO (51.23%) [15] and could increase plant uptake of P. Combining Si and microorganisms application has been proposed to effectively induce and improve plant growth and nutrition [16–18]. It was earlier observed that AMF and Si (from SiO_2—a constituent of RP) work together to improve plant growth regardless of stress conditions. Similarly, PBS and SiO_2 synergistically help plants better uptake P [19, 20]. Another study indicates that SiO_2 contained in the rock phosphate is a quasi-essential nutrient and is beneficial to plants, especially when under different stresses such as drought, heavy metal toxicity, nutritional imbalance, plant pathogens, and salinity [21]. Si fertilization increases P levels in different crops and improves plant growth by enhancing P availability for plants [22]. PBS generally have the ability to weather silicates, likely because basic metabolic activities such as organic acids production and respiration can cause the weathering of minerals [23].

3. Phosphate solubilization sources and mechanism of action

Phosphate-solubilizing microorganisms technology improves saline-alkaline soil fertility and agricultural use without causing any environmental or deleterious health implication that accompanies the continuous use of synthetic fertilizers [24]. The phenomenon of P desorption by PSM usually occurs along with the drop of pH [25]. The ability of PSM to convert insoluble organic and inorganic phosphorus depends on the nutritional richness of the soil and the physiological and growth status of the organism. PSM isolated from saline-alkaline soils–soils with a high level of nutrient deficiency, or soils from extreme temperature environments, have the tendency to solubilize more phosphate than PSM of soils from moderate conditions [14].

P solubilization has been reported to be secretion of organic acids (e.g., oxalic acid, citric acid, acetic acid, succinic acid, etc.) produced by phosphate-solubilizing microbes to solubilize insoluble P by lowering of pH. The hydroxyl and carboxyl groups present in the organic acids chelating the cations (mainly calcium) bound to phosphate, leading to increased solubility and availability of mineral phosphates, to ensure the microbial growth, organic acid production, and RP solubilization of metabolizable carbon compounds [26–30].

Mechanism of action of P solubilization could be observed when a phosphate-solubilizing microbe is grown in a specialized highly alkaline medium known as Pikovskaya medium. These phosphate-solubilizing microbes are then incubated at 30°C for six days. There has been a conflicting report on the influence of temperature on phosphorus solubilization by microbes. White et al. [31] found 20–25°C as the optimum temperature for maximum microbial phosphorus solubilization while 28°C was reported by [32, 33]. In addition, others including [34–37] have recorded 30°C as the best temperature for P solubilization. [36, 38] reported P solubilization at extreme temperature of 45°C in desert soil while [39] reported solubilization at a low temperature of 10°C. The ability of these phosphate-solubilizing microbes to form halo zones when cultured on the PVK medium is a clear indication that the organism can solubilize phosphate. Later, the PSB/AMF could then be centrifuged at 5000 rpm for 10 min [40]. Supernatant of blended cultures was filtered through 0.20-μm syringe filters (cellulose acetate), and the organic acid could then be detected using HPLC (shimadzu SPD-M20A) with PDA detector by using methanol 5% and 0.1% TFA 95% as mobile phase with flow rate of 1 ml per minute with column particle size 5 μm, length 25 cm, and column diameter 4.6 mm (Merck C18 column). The extent of P solubilization by the fungal strain could then be determined by measuring the clear zone (in mm) around the colonies by taking the halo zone formation into consideration [25, 41–44].

4. Role of rock phosphates and phosphate-solubilizing microbes in food security

Food security is heavily dependent on phosphorus availability—an essential component of the mined phosphate rock [45]. The fertilizer industry is aware of the fact that the quality of RP reserve is declining and the extraction, processing, and shipping of phosphate rock are highly expensive [46–48]. With the current global population boom, coupled with insecurity challenges in mines posed by armed banditry herdsmen, Boko haram, and the Islamic State in West Africa Province (ISWAP) in this part of the world, optimizing food security is essential in curbing poverty and hunger management.

Emphasis on this chapter is not limited to improving RP solubility alone but also how to ensure food security under diminishing P resources due to injudicious use of P fertilizers [47]. This is made possible by application of phosphate-solubilizing microbes/biofertilzers in addition to the RP. This could be a more promising approach for safer and sustainable agriculture. PSM include bacteria (*Pseudomonas* sp., *Agrobacterium* sp., *Bacillus* sp., *Actinobacteria* etc), fungi (*Achrothcium*, *Alternaria*, *Arthrobotrys*, *Aspergillus*, etc.), and Arbuscular mycorrhizae (*Funneliformis mosseae, Rhizophagus irregularis,* etc). [12, 17, 48].

Therefore, inoculation of soil with PSMs was reported to be a widely accepted environmentally friendly approach for increasing soil soluble P concentrations and agricultural productivity [48]. It was established that bioavailable P content in soil is an important element that improves plant P uptake, which in turn results in higher crop yields [14, 24]; most studies have considered PSM as a promising inoculant/ biofertilizer for raising the productivity of agronomic crops in agroecological niches [49–51]. Soil microorganisms improve plant nutrient acquisition. They are involved in an extensive range of biological processes as well as the conversion of insoluble soil nutrients [34]. A number of these soil microbes are capable of solubilizing and mineralizing insoluble soil phosphorus for the growth of plants. To this end, application of phosphate PSMs as biofertilizers will bring favorable effects on plant growth and reduce the cost of production, which may likely enhance the farmers' income in sub-Saharan Africa.

5. Conclusion

Based on previous studies, we could conclude that AMF, PBS, and RP consortium could synergistically act to more effectively increase the plant uptake of P, improving the growth of different plant more than when each was solely applied as soil amendment. These in turn boost yield and generate income of poor Sahelian subsistent farmers. In order to avoid a future food-related crisis, phosphorus scarcity needs to be recognized and addressed in contemporary discussions on global environmental change and food security. This review has shown that phosphate rocks and phosphate-solubilizing microbes have great potential as biofertilizers, improving soil inorganic phosphate levels and increasing its bioavailability for plant use. This will no doubt promote sustainable agriculture, improve the fertility of the soil, and hence increase crop productivity.

6. Future prospects

There is great potential in formulation of biofertilizer using RP and phosphate-solubilizing microbe. This could potentially overcome abiotic stress such as drought and salinity, which could improve crop yield. AMF, PSB, and RP interaction to affect P availability is poorly understood [35], thus a better understanding of AMF and PSB and RP would allow growers to rely less on chemical P fertilizers and instead utilize biological processes to maintain fertility and enhance plant growth. AMF and PBS could work together to yield sustainable plant growth in malnourished environments. Combinations of AMF and PSB are commonly used to increase crop yields, improve fruits quality, boost phytoremediation, enhance the fertilizer nutrient use efficiency, lower chemical fertilization application requirements, and increase salinity tolerance [52, 53]. The use of silicon (Si) fertilizer has also been proposed as an environmentally friendly, ecologically compatible method of improving plant growth and the resistance to multiple environmental stresses including nutritional imbalances. Previous studies have reported that Si increases plant uptake of P [21, 54, 55]. Combining Si and microorganism applications has been proposed to efficiently improve plant growth and nourishment. Previous studies have observed that AMF, PBS, R, and Si could work together to improve plant growth regardless of the stress conditions [56–61].

Conflict of interest

The authors declare that they have no known competing financial or personal relationship that could have appeared to influence the report in this paper.

Author details

Zainab Muhammad Bello[1*], Sanusi Muhammad[1], Adamu Aliyu Aliero[1],
Adamou Rabani[2] and Ibrahim Aliyu Dabai[3]

1 Department of Biological Sciences, Usmanu Danfodiyo University, Sokoto, Nigeria

2 West African Science Service Center on Climate Change and Adapted Land Use, Abdou Moumouni University (UAM), Niamey, Niger

3 Department of Microbiology, Usmanu Danfodiyo University, Sokoto, Nigeria

*Address all correspondence to: zainab.bello@udusok.edu.ng

References

[1] Steen I. Phosphorus availability in the 21st century: Management of a nonrenewable resource. Phosphorus and Potassium. 1998;**217**:25-31

[2] Gunther F. A Solution to the Heap Problem: The Doubly Balanced Agriculture: Integration with Population. 2005. Available from: http://www.holon.se/folke/kurs/Distans/Ekofys/Recirk/Eng/balanced.shtml

[3] Fernandes M, Bastida F, Jehmlich N, Martinovic T, Vetrovsky T, Baldrian P, et al. Functional soil mycobiome across ecosystems. Journal of Proteomics. 2021;**252**:104428. DOI: 10.1016/j.jprot.2021.104428

[4] Cornell C, Kokkoris V, Turcu B, Dettman J, Stefani F, Corradi N. The arbuscular mycorrhizal fungus Rhizophagus irregularis harmonizes nuclear dynamics in the presence of distinct abiotic factors. Fungal Genetics and Biology. 2022;**158**:103639. DOI: 10.1016/j.fgb.2021.103639

[5] Espiritu K. How to Use Rock Phosphate Fertilizer In your Garden, California; 2022. Available from: https://www.epicgardening.com. [Accessed: June 22, 2022]

[6] Omar SA. The role of rock-phosphate-solubilizing fungi and vesicular–arbusular-mycorrhiza (VAM) in growth of wheat plants fertilized with rock phosphate. World Journal of Microbiology and Biotechnology. 1998;**14**(2):211-218

[7] Das D, Paries M, Hobecker K, Gigl M, Dawid C, Lam HM, et al. Phosphate starvation response enables arbuscular mycorrhiza symbiosis. bioRxiv. 2021.11.05.467437. DOI: 10.1101/2021.11.05.467437 Now published in *Nature Communications* DOI: 10.1038/s41467-022-27976-8

[8] Chagnon P-L, Bradley R, Lafond J, Paré M, Penaud V. Trait-based and phylogenetic filtering of arbuscular mycorrhizal fungal communities under long-term agricultural practices. Plant and Soil. 2022;**471**:1-2. DOI: 10.1007/s11104-021-05155-w

[9] Sayeda S, Ibrar M, Hadi F. Effect of arbuscular mycorrhizal Fungi (AMF) on fatty acid contents of selected sunflower (Helianthus annuus L.) hybrids under various levels of rock phosphate. American-Eurasian Journal of Agriculture and Environment Science. 2016;**16**(4):719-728. DOI: 10.5829/idosi.aejaes.2016.16.4.12927

[10] Kobae Y. Dynamic phosphate uptake in arbuscular mycorrhizal roots under field conditions. Frontiers in Environmental Science. 2019;**6**:159. DOI: 10.3389/fenvs.2018.00159

[11] Papadopoulos G, Chouliaras N, Jacquin F. Interference between soil microflora activity and natural phosphates availabilty. Agriculural Reseach, Issued by the ministry of Agriculture, Greek Government. 1986;**10**:231-241

[12] Lasudee K, Tokuyama S, Lumyong S, Pathom-Aree W. Actinobacteria associated with arbuscular mycorrhizal *Funneliformis mosseae* spores, taxonomic characterization and their beneficial traits to plants: Evidence obtained from mung bean (*Vigna radiata*) and Thai jasmine Rice (*Oryza sativa*). Frontiers in Microbiology. 2018;**9**:1247. DOI: 10.3389/fmicb.2018.01247

[13] Etesami H, Jeong BR, Glick BR. Contribution of arbuscular mycorrhizal

Fungi, phosphate-solubilizing Bacteria, and silicon to P uptake by plant. Frontiers in Plant Science. 2021;**12**:699618. DOI: 10.3389/fpls.2021.699618. PMID: 34276750; PMCID: PMC8280758

[14] Alori ET, Glick BR, Babalola OO. Microbial phosphorus solubilization and its potential for use in sustainable agriculture. Frontiers in Microbiology. 2017;**8**:971. DOI: 10.3389/fmicb.2017.00971

[15] Ai-Falih AM. Effect of silicon compounds on oligotrophic soil bacteria. Saudi. J. Biol. Sci. 2003;**10**:131-137

[16] Ahemad M, Kibret M. Mechanisms and applications of plant growth promoting rhizobacteria: Current perspective. J. King Saud University-science. 2014;**26**:1-20. DOI: 10.1016/j.jksus.2013.05.001

[17] Anda CCO, Opfergelt S, Declerck S. Silicon acquisition by bananas (cV Grande Naine) is increased in presence of the arbuscular mycorrhizal fungus Rhizophagus irregularis MUCL 41833. Plant and Soil. 2016;**409**:77-85. DOI: 10.1007/s11104-016-2954-6

[18] Appanna V. Efficacy of phosphate solubilizing bacteria isolated from vertisols on growth and yield parameters of sorghum. Research Journal of Microbiology. 2007;**2**:550-559. DOI: 10.3923/jm.2007.550.559

[19] Bijalwan P, Jeddi K, Saini I, Sharma M, Kaushik P, Hessini K. Mitigation of saline conditions in watermelon with mycorrhiza and silicon application. Saudi Journal of Biological Sciences. 2021;**28**(7): 3678-3684, ISSN 1319-562X. DOI: 10.1016/j.sjbs.2021.05.019

[20] Abbas T, Balal RM, Shahid MA, Pervez MA, Ayyub CM, Aqueel MA, et al. Javaid silicon-induced alleviation of NaCl toxicity in okra (Abelmoschus esculentus) is associated with enhanced photosynthesis, osmoprotectants and antioxidant metabolism. Acta Physiologiae Plantarum. 2015;**37**:6. DOI: 10.1007/s11738-014-1768

[21] Zarei M, Saleh-Rastin N, Alikhani HA, Aliasgharzadeh N. Responses of lentil to co-inoculation with phosphate-solubilizing rhizobial strains and arbuscular mycorrhizal fungi. J. Plant Nutrition. 2006;**29**:1509-1522. DOI: 10.1080/01904160600837667

[22] Etesami H, Jeong BR. Silicon (Si): Review and future prospects on the action mechanisms in alleviating biotic and abiotic stresses in plants. Ecotoxicology and Environmental Safety. 2018;**147**:881-896. DOI: 10.1016/j.ecoenv.2017.09.063

[23] Kostic L, Nikolic N, Bosnic D, Samardzic J, Nikolic M. Silicon increases phosphorus (P) uptake by wheat under low P acid soil conditions. Plant and Soil. 2017;**419**:447-455. DOI: 10.1007/s11104-017-3364-0

[24] Zhu F, Qu L, Hong X, Sun X. Isolation and characterization of a phosphate solubilizing halophilic bacterium Kushneria sp. YCWA18 from Daqiao Saltern on the coast of yellow sea of China. Evidence-based Complementary and Alternative Medicine. 2011;**2011**:615032. DOI: 10.1155/2011/615032Z

[25] Bandeppa GS, Kumar K, Pc L, Manjusha PGS, Phule A, Chandrakala C. Phosphate-solubilizing microbial inoculants for sustainable agriculture. Trends of Applied Microbiology for Sustainable Economy. 2022:501-524. DOI: 10.1016/B978-0-323-91595-3.00010-0

[26] Mander C, Wakelin S, Young S, Condron L, O'Callaghan M. Incidence and diversity of phosphate-solubilising

bacteria are linked to phosphorus status in grassland soils. Soil Biology and Biochemistry. 2012;**44**(1):93-101. DOI: 10.1016/j.soilbio.2011.09.009

[27] Khan MS, Zaidi A, Wani PA. Role of phosphate-solubilizing microorganisms in sustainable agriculture – A review. Agronomy for Sustainable Development. 2007;**27**(1):29-43. DOI: 10.1051/agro:2006011

[28] Trabelsi D, Cherni A, Zineb AB, Dhane SF, Mhamdi R. Fertilization of *Phaseolus vulgaris* with the Tunisian rock phosphate affects richness and structure of rhizosphere bacterial communities. Applied Soil Ecology. 2017;**114**:1-8. DOI: 10.1016/j.apsoil.2016.11.014

[29] Wei Y, Zhao Y, Wang H, Lu Q, Cao Z, Cui H, et al. An optimized regulating method for composting phosphorus fractions transformation based on biochar addition and phosphate-solubilizing bacteria inoculation. Bioresource Technology. 2016;**221**:139-146. DOI: 10.1016/j.biortech.2016.09.038

[30] Reyes I, Valery A, Valduz Z. Phosphate-solubilizing microorganisms isolated from rhizospheric and bulk soils of colonizer plants at an abandoned rock phosphate mine. Plant and Soil. 2006;**287**(1-2):69-75. DOI: 10.1007/s11104-006-9061-z

[31] White C, Sayer JA, Gadd GM. Microbial solubilization and immobilization of toxic metals: Key biogeochemical processes for treatment of contamination. FEMS Microbiology Reviews. 1997;**20**:503-516. DOI: 10.1111/j.1574-6976.1997.tb00333.x

[32] Kang SC, Ha GC, Lee TG, Maheshwari DK. Solubilization of insoluble inorganic phosphates by a soil inhabiting fungus sp. Ps 102. Current Science. 2002;**79**:439-442

[33] Varsha NHH. *Aspergillus aculeatus* as a rock phosphate solubilizer. Soil Biology and Biochemistry. 2002;**32**:559-565

[34] Kim KY, Jordan D, Kirshanan HB. *Rahnella aquatilis*, a bacterium isolated from soybean rhizosphere, can solubilize hydroxyapatite. FEMS Micobiol. Lett. 1997a;**153**:273-277. DOI: 10.1016/S0378-1097(97)00246-2

[35] Rosado AS, De Azevedo FS, da Croz DW, Van Elas JD, Seldin L. Phenotypic and genetic diversity of *Paenibacillus azotofixans* strains isolated from the rhizosphere soil of different grasses. Journal of Applied Microbiology. 1998;**84**:216-226. DOI: 10.1046/j.1365-2672.1998.00332.x

[36] Johri JK, Surange S, Nautiyal CS. Occurrence of salt, pH and temperature tolerant phosphate solubilizing bacteria in alkaline soils. Current Microbiology. 1999;**39**:89-93. DOI: 10.1007/s002849900424

[37] Fasim F, Ahmed N, Parsons R, Gadd GM. Solubilization of zinc salts by bacterium isolated by the air environment of tannery. FEMS Microbiology Letters. 2002;**213**:1-6. DOI: 10.1111/j.1574-6968.2002.tb11277.x

[38] Nahas E. Factors determining rock phosphate solubilization by microorganisms isolated from soil. World Journal of Microbiology and Biotechnology. 1996;**12**:567-572. DOI: 10.1007/BF00327716

[39] Nautiyal CS, Bhadauria S, Kumar P, Lal H, Mondal R, Verma D. Stress induced phosphate solubilization in bacteria isolated from alkaline soils. FEMS Microbiology Letters. 2000;**182**:291-296. DOI: 10.1111/j.1574-6968.2000.tb08910.x

[40] Cordell D, Jan-Olof D, Stuart W. The story of phosphorus: Global food

security and food for thought. Global Environmental Change. 2009, 2009;**19**(2):292-305, ISSN 0959-3780. DOI: 10.1016/j.gloenvcha.2008.10.009

[41] Runge-Metzger A. Closing the cycle: Obstacles to efficient P management for improved global food security. In: SCOPE 54 – Phosphorus in the Global Environment – Transfers, Cycles and Management. 1995

[42] Driver J. Phosphates recovery for recycling from sewage and animal waste. Phosphorus and Potassium. 1998;**216**:17-21

[43] Smil V. Phosphorus in the environment: Natural flows and human interferences. Annual Review of Energy and the Environment. 2000b;**25**:53-88

[44] Tian J, Ge F, Zhang D, Deng S, Liu X. Roles of phosphate solubilizing microorganisms from managing soil phosphorus deficiency to mediating biogeochemical P cycle. Biology. 2021;**10**:158. DOI: 10.3390/biology10020158

[45] Zaidi A, Ahemad M, Oves M, Ahmad E, Khan MS. Role of phosphate-solubilizing Bacteria in legume improvement. In: Microbes for Legume Improvement. Cham, Switzerland: Springer; 2010. pp. 273-292

[46] Mendes GDO, Galvez A, Vassileva M, Vassilev N. Fermentation liquid containing microbially solubilized P significantly improved plant growth and P uptake in both soil and soilless experiments. Applied Soil Ecology. 2017;**117**:208-211

[47] Ahemad M. Phosphate-solubilizing bacteria-assisted phytoremediation of metalliferous soils: A review. 3. Biotech. 2015;**5**:111-121

[48] Babalola OO, Glick BR. Indigenous African agriculture and plant associated microbes: Current practice and future transgenic prospects. Scientific Research and Essays. 2012a;**7**:2431-2439

[49] Sagervanshi A, Kumari P, Nagee A, Kumar A. Isolation and characterization of phosphate Solublizing Bacteria from Anand agriculture soil. International Journal of Life Science and Pharma Research. 2012;**2**:256-266

[50] Nautiyal CS. An efficient microbiological growth medium for screening phosphate solubilizing microorganisms. FEMS Microbiology Letters. 1999;**170**(1):265-270. DOI: 10.1111/j.1574-6968.1999.tb13383.x

[51] Nagar ACNV. Media optimization for inorganic phosphate solubilizing bacteria isolated from Anand agriculture soil. International Journal of Life Science & Pharma Research. 2012;**2**:245-255

[52] Fasim F, Ahmed N, Parsons R, Gadd GM. Solubilization of zinc salts by a bacterium isolated from the air environment of a tannery. FEMS Microbiology Letters. 2002;**213**(1):1-6. DOI: 10.1111/j.1574-6968.2002.tb11277.x

[53] Etesami H, Noori F. Soil salinity as a challenge for sustainable agriculture and bacterial-mediated alleviation of salinity stress in crop plants. In: Kumar M, Etesami H, Kumar V, editors. Saline Soil-Based Agriculture by Halotolerant Microorganisms. 2019

[54] Lee Y, Krishnamoorthy R, Selvakumar G, Kim K, Sa T. Alleviation of salt stress in maize plant by co-inoculation of arbuscular mycorrhizal fungi and *Methylobacterium oryzae* CBMB20. Journal of Korean Society for Applied Biological Chemistry. 2015;**58**:533-540. DOI: 10.1007/s13765-015-0072-4

[55] Ordookhani K, Khavazi K, Moezzi A, Rejali F. Influence of PGPR and AMF

on antioxidant activity, lycopene and potassium contents in tomato. African J. Agricul. Res. 2010;**5**:1108-1116

[56] Nadeem SM, Ahmad M, Zahir ZA, Javaid A, Ashraf M. The role of mycorrhizae and plant growth promoting rhizobacteria (PGPR) in improving crop productivity under stressful environments. Biotechnology Advances. 2014;**32**:429-448. DOI: 10.1016/j.biotechadv.2013.12.005

[57] Sharma SB, Sayyed RZ, Trivedi MH, Gobi TA. Phosphate solubilizing microbes: Sustainable approach for managing phosphorus deficiency in agricultural soils. Springerplus. 2013;**2**:587

[58] Hajiboland R, Moradtalab N, Aliasgharzad N, Eshaghi Z, Feizy J. Silicon influences growth and mycorrhizal responsiveness in strawberry plants. Physiology and Molecular Biology of Plants. 2018;**24**:1103-1115. DOI: 10.1007/s12298-018-0533-4

[59] Etesami H. Can interaction between silicon and plant growth promoting rhizobacteria benefit in alleviating abiotic and biotic stresses in crop plants? Agricul. Ecosystems Environ. 2018;**253**:98-112. DOI: 10.1016/j.agee.2017.11.007

[60] Moradtalab N, Hajiboland R, Aliasgharzad N, Hartmann TE, Neumann G. Silicon and the association with an arbuscular-mycorrhizal fungus (*Rhizophagus clarus*) mitigate the adverse effects of drought stress on strawberry. Agronomy. 2019;**9**:41. DOI: 10.3390/agronomy9010041

[61] Rezakhani L, Motesharezadeh B, Tehrani MM, Etesami H, Hosseini HM. Effect of silicon and phosphate-solubilizing bacteria on improved phosphorus (P) uptake is not specific to insoluble P-fertilized sorghum (*sorghum bicolor* L.) plants. J. Plant Growth Regulat. 2019a;**39**:239-253. DOI: 10.1007/s00344-019-09978-x

Chapter 6

Taking Stock of Local Land Rush and Their Development Benefits for Women Farmers and Rural Livelihoods in Cameroon

Christiana V. Abonge

Abstract

Land is a major resource for rural women who are largely dependent on agriculture as a major source of livelihood. The rise in large scale land acquisition by local investors has made land scarcer, affecting the livelihoods of women and the rural population. Because the poor and women are less likely to own land independently, they are more vulnerable and become landless with limited livelihood options and bear disproportionate costs of land deals. The study intends through the use of a survey, interviews and focus group discussions, to investigate the effects of local large scale land deals on women's livelihoods in North Western Cameroon and how this affects household food security in these communities. Drawing on empirical evidence from the study, women who are mainly responsible for subsistence food crop production, have very limited access to and ownership rights to land. While local investors promise employment opportunities and local development in rural areas, women are also particularly deprived of farmland, which is a major source of livelihood and household subsistence. Thus, local land acquisition fails to trickle down to social and infrastructural development and is more likely to restrict women's ability to meet the basic food needs of their households.

Keywords: rural livelihoods, local investors, social development, women farmers, Cameroon farmers

1. Introduction

The dearth of employment and industrial activity in most rural areas make agriculture a predominant economic activity particularly for rural peoples, especially women and the poor. Women in Cameroon provide the backbone of the agricultural sector as a greater majority of economically active female labour force is employed in agriculture [1–6]. Extensive evidence shows that rural women in Cameroon make up a majority of the active labour force in the food crop sector producing about 90% of food crops in the country [7, 8]. As food crop producers and processors, women play a crucial role in ensuring family survival and safeguarding national food security. Women grow food to feed their families, making subsistence food production one of the major agricultural and livelihood activities of women. While men may migrate to urban areas in search of work and other income earning opportunities, women who are compelled to be heads of households rely on subsistence food crop production for household survival [9].

The place of women's subsistence agricultural production also calls to mind the significance of women's access, use and control over production resources particularly land, for agricultural production. This has met with increased efforts to enhance small holder (women) farmer's access to and control over land. The efforts to increase subsistence agricultural productivity however fall short of meeting the growing needs and demands of agricultural producers in general and rural and female subsistence producers in particular. Despite the predominant role of women in food crop production, they have limited access to and control over land. They lack ownership and inheritance rights though they can acquire use rights to cultivable land through male relatives. This limitation of increasing women's access to and control over land is exacerbated by land grabbing that has been observed in parts of Sub-Saharan Africa including Cameroon [10]. The social construction of gender that gives women traditional roles within African families as house-keepers only deprive women's access and control over community land making them more vulnerable and most affected by land grabbing [11–13].

In the process, women and the poor are more vulnerable and become landless with limited livelihood options and bear disproportionate costs of large-scale land deals. While displacement of community members often ignores women's role of subsistence food crop production, the process also increases pressure on women's tenure. On the other hand, women are often left out and are not equally represented in decisions concerning land transactions; either because of women's low levels of education and their lack of information for any informed decision or their lack of legal resources to defend land claims [14, 15]. Also, women are usually marginalized and ignored with respect to the sharing of benefits associated to large scale land deals. Benefits, which can either be in the form of compensation, income generation opportunities or employment, target title holders who are usually men, hence leaving out women who may only benefit as casual labourers in precarious part time work with low wages.

Despite women's role in food crop production and household subsistence, customary and cultural limitations and negative practices related to property ownership and equal rights to land most often hinder women's agrarian activities [16]. Similarly, other factors including growing large scale land acquisition (LSLA) also tend to make land a very scarce resource particularly for women and the poor. The LSLA phenomenon, which involves different groups of actors (investors/MNC/TNC, state/governments, communities and community leaders, men, women, etc) has greatly affected the livelihoods of the populations involved.

It has been observed that the effects of these land deals are different on different communities as well as different members of the community. The process usually has differential impacts on members of the community and while some may benefit from the process, it is likely that others do lose. Most often, ignoring the gender impacts of large-scale land transactions is a common practice that overlooks the impact of land deals on women and other vulnerable members of the community. The process and involvement in LSLA including the impact and benefit of these transactions is defined by several factors including how land is acquired and tenure rights, the composition of the teams responsible for or executing the transaction, and the impact of land deals on community members [17–19].

Another major criticism of large scale land deals is the issue of poor land governance and corrupt and unaccountable decision making including the weak protection and sidelining of the poor and vulnerable members of the community. Research on LSLA confirms that decisions on land transactions rarely involve those who use the land but are made by and large by community leaders including state and government officials [16, 20]. The process of decision making related to land deals consequently raises concerns about the impact (gendered) of land acquisition processes on the livelihood options of members of the community in general and women in particular.

2. Land tenure in Cameroon

Access to land can in part be defined by customary land rights and practices which are not often secured by law [18, 21]. Rooted in the colonial legacy that enhanced patriarchal norms and structures, land management structures in Cameroon in general and the North West Region in particular usually render women's access to land difficult by promoting men's access and control over land. This also defines the right to control and use land giving men more power and authority over land than women [22, 23]. Understanding women's access to and control over land requires an analysis of various land tenure systems in general and women's land tenure patterns in particular.

Land tenure according to [24] refers to the terms and conditions associated with holding, using and transacting land, while land rights refer to the right to occupy and use land for various purposes with the ability to make improvements on the piece of land. Land rights also give owners the right to give, lease, rent and/or donate a piece of land [19–21]. This indicates that having more rights to land can be associated with more sustainable livelihoods for those with more rights to land than the landless and those with limited rights to land. In addition to helping land owners build an asset base that may lead to sustainable livelihood, having land rights may provide a source of power, prestige and a basis to recover from shocks and stress, as well as a means of economic production and other activities for a living [14, 19, 21, 24]. Land rights for women increase their bargaining power within the household with a resulting increase to household welfare [15]. Thus, secure land rights are very instrumental to economic development and a catalyst for economic activity.

The formal and customary land tenure systems have been identified as two main frameworks and/or institutions governing ownership and access to land [15, 18, 21]. It is important to note that because of the importance attached to customs, civil law protections are often not respected. Even though many laws recognize and protect the rights of women to land (property and land ownership rights, equal inheritance rights and marital property rights), enforcement of these laws is quite irregular and difficult. For example, Cameroon's 1996 Constitution gives all citizens the right to own property individually or in association with others, where ownership is defined by the right to use, enjoy and transfer property. On the other hand local restrictions and customary practices against women's ownership of land most often supersedes written law and therefore determine which rights to land can or cannot be exercised. The difficulty in implementing these laws could be as a result of conflicting legislation, institutional weaknesses and gender bias. The interest of this research is to note the convergence of customary laws and practices, institutions and traditional structures and formal laws to entrench and reinforce the already unequal balance of gender power relations and gender inequality within the community.

As observed by [25], land in Cameroon can be considered as a community resource in terms of space inherited from the ancestors. As a community resource, land is managed by community representatives who can either be chiefs or traditional rulers and it is commonplace for land (community) to be inherited and transmitted from one generation to the other. Based on the pre-colonial conception of the ownership of means of production, land was conceived during this period as collective property for collective use and management. This however changed with the advent of colonialism that adopted a Eurocentric view of land as individual resource and thus the infiltration of individual property rights, highlighting the colonialist intension to control and manage land. Thus, it is important to note that the Cameroonian land law is a derivative of this European principle of individual ownership of land [21]. Based on this principle, the land ordinance of 1974 supports and promotes private ownership and the commercialization of land. According to this law, it is imperative for privately owned land to be registered and titled, without which such untitled and unregistered land is classified as national land [25, 26]. It is only when land is registered and titled that the land holder can be associated with ownership rights Rural communities can also title their customary land, and titles are the only protection against state appropriation. Given the difficult, long and expensive process registering land, it is difficult for customary land owners to register and secure communal land. Thus, such communities enjoy only usufruct rights to their land. This reinforces insecure tenure for most communities and individuals as a greater majority of farmland and communal land held under customary law is classified as national land. Generally, while land in rural areas is predominantly subject to customary law where traditional leaders (Fons, chiefs or Lamidos) administer and manage these lands, individual families may receive the rights to use land which are heritable through male kin [19, 21, 26, 27].

Cameroon's land tenure regime is therefore based on this and subsequent laws, emphasizing land registration and titling. Nonetheless, most land is still managed informally through local arrangements, whose flexible rules may create uncertainty, foster land conflicts, and hamper local development. Even though land reforms in 2005 simplified land titling by reducing the number of steps and departments involved, and cut the time needed to obtain a land title from several years to less than one, it failed to address significant hurdles including contradictory laws, poor record keeping, and reliance on traditional authorities to allocate land rights [19, 21].

Land registration seems to be more of an urban phenomenon given that rural communal land is more agricultural and is hardly registered. Notwithstanding, an estimated 80% of plots in urban and peri-urban areas are held under customary law. Plots are obtained through renting or purchase on the informal market [28–31]. The gaps in the land administration processes and infrastructure account in part to the slow transition from customary property rights to private ownership of property, increasing rather than diminishing land tenure insecurity. Cases of multiple sales of the same land, false land certificates, and inaccuracies in boundary definitions are commonplace, giving rise to frequent conflicts and disputes, also demonstrating that the registration process does not improve tenure security.

3. Large scale land deals and the socio-cultural context in the north west region

Large scale land deals commonly referred to as land grabs is characterized by the acquisition of large portions of land mostly by foreign and or local investors. These land deals are often negotiated and assisted by national elites and governments on the one hand and local councils or elites longing for foreign investment [16, 18]. While investors may consider land grabbing as a means of providing development to the locals in terms of better and improved access to better roads and social services such as health centres, access to potable water and electricity, such land deals are characterized by the violation of human rights, are not based on free, prior and informed consent of the land users, and usually ignore the social, economic and environmental impacts including the way the process is gendered [16, 22]. There is increasing concern about the positive and negative impacts of these land deals on the livelihoods of men and women who are those who are usually displaced.

The North West Region (NWR) is characterized by various ethnic groups reflecting widespread ethnic diversity also observed in the country [6, 32]. The major ethnic groups (otherwise considered indigenes) in the province include the Bali, Widekum, Nso, Mbembe and the Aghem. Another population group (minority) in the province constituting about 5–10% of the total population is the Mbororo; a group of pastoralists and Muslim migrants from Northern Cameroon. Deriving from the ethnic diversity is a number of centralized villages and chiefdoms, which are autonomous with similar economic and socio-political structures and are by and large culturally and linguistically heterogeneous [6, 33, 34].

Although there are various forms of religious practices (Christianity, Islam, and traditional beliefs), the people of the NWR are predominantly Christians with the minority Mbororo mainly Muslims [33, 34]. The diverse ethnic background is characteristic of variations in traditional norms and practices (purdah and female seclusion, age and forms of marriage, etc). Whilst adherence to these practices may also define the extent of gender inequality, which does vary by ethnic group, purdah and female seclusion is common amongst the Mbororo and Muslims [32, 33]. There are plausible ways through which these practices can influence women's position within the household. Pelican [32] suggests an association between polygamy, religious practice and ethnicity. Polygamy, which is widespread amongst Muslims and those observing traditional religious practices (than amongst Christians), reinforces negative social standing and limited access to resources for women [6, 32, 33]. For instance, male financial support may be limited in polygamous homes due to greater competition for resources [32], exacerbating the burden of household subsistence on women and greater constraints with access to health care, education and other basic needs for children in such homes. Whilst this may raise concerns as to the changing position of women, it also highlights the role of socio-cultural factors in defining the availability and accessibility of production resources including land to women within the household.

4. Assessing the benefits and impact of land deals: the need and the challenge

Large scale land deals are usually associated with positive development justifying the positive response and acceptance from local populations. Multinational Corporations (MNCs) and local investors involved in the process make promises to local community members in the form of providing social amenities and infra-structure (health centers, schools, roads and bridges, water and electricity and employment) as benefits that will accrue to local communities in exchange for their land [17]. The multi-national companies involved and the state define access to these social services and amenities as positive development that justifies the displacement of community members from cultivable land and livelihood sources. This suggests that while most

MNCs and local investors value and emphasize on meeting their goals and objectives and their own development, they fail to recognize the capability and livelihood security of the people [17, 22]. There is evidence that the poor and rural population in general and women in particular are deprived of their assets, capabilities and activities that provided them with a source of living. There have been diverse implications of land deals on community members registered in different localities ranging from land access, income generating and livelihood options. The nexus between land tenure rights, land grabbing and food insecurity is very prominent and critical in achieving food security in Cameroon. While insecure tenure has attracted the rampant acquisition of large-scale land by foreign and local investors, this has also contributed to the scarcity of farmlands and the shortage of land for indigenous people destroying the natural ecosystem that the local people depended on for their livelihood and also affecting the availability of food crops.

5. Methodology and area of study

The research is descriptive and exploratory and targets women farmers and households in selected communities in the region that have been affected by large scale land deals. It draws on field research carried out in the North West Region (NWR) of Cameroon and employed a quantitative methodological approach that provided an in-depth understanding of the experiences of members of the community studied. The study examines underlying connections between large scale land deals and livelihood activities in general and women's livelihoods in particular including the link to household food security. Data for this study was gathered using both primary and secondary sources. Primary data, which constituted the principal source of information for the study was elicited using multiple data collection methods. Through a survey and in-depth interviews, the study provides an account of respondents' views and perceptions of the effects of local large-scale land deals on rural livelihood options by exploring the meanings women themselves and the rural population give to these issues. Information was also sought from respondents on the various actors involved in the process and the effects of land deals on rural livelihoods and household food security. Meanwhile, the survey and interview also collected data on the gender roles and livelihood options or choices of respondents, the land tenure arrangements in the community and how this affects the access and control profile of men and women with respect to land in the selected communities. Data generated for the study was analyzed using themes identified both from literature and interviews, which were also guided and captured by the objectives of the study.

6. Findings

6.1 Women as subsistence food crop farmers

As confirmed by the study, subsistence agriculture remains a major source of employment and economic activity for the population of the North West Region especially women. According to field evidence, an overwhelming majority of rural women in the region are farmers even though there are some women who diversify their income earning opportunities by combining subsistence agriculture with informal business activities as a livelihood option. The dearth of industrial activity in the region

Age of women		
Age (years)	**Frequency**	**Percent**
≤20	0	0
21–30	19	8.3
31–40	30	13.04
41–50	107	46.6
51–60	51	22.1
Above 60	23	8.7
Total	230	100
Marital Status	Frequency	Percentage
Married	147	63.9
Singe	31	13.5
Widow	45	19.6
Divorced	7	3
Total	230	100
Level of Education		
No formal Education	61	26.5
Primary	102	44.3
Secondary	58	25.2
High School	09	3.09
Total	230	100
Occupation		
Farming	217	94.3
Business	10	4.4
Others	3	1.4

Source: Generated from field data.

Table 1.
Demographic characteristics of women in the study area.

makes agriculture and self-employment predominant economic activities particularly for the rural population, women and the poor. In the region and the country in general, women provide the backbone to the agricultural sector as a greater majority of economically active female labour force is employed in agriculture (**Table 1**).

While most of the women who engaged in the study (46.6%) are between 41 and 50 years old, it was observed that close to three quarters (70.8%) of the women have only primary education. This confirms the predominance of women in the agricultural sector that requires little or no education and any specialized skills for employment in the sector in general and the subsistence food crop sector in particular. Meanwhile, about two thirds (63.9%) of the study population are married suggesting that this population are likely to have use rights to land through their spouses.

Women in Cameroon can be linked with various socio-economic roles, which cut across a wide range of activities including farming, trading and working in paid labour force. They make up about 38.2% of the total labour force in the country and are a significant part of the labour force in the agricultural sector playing indispensable roles in the production, processing, and marketing of food crops [6, 35, 36]. Approximately 83% of economically active female labour force is employed in agriculture. Extensive evidence observes that women in rural areas make up a majority of

Crops cultivated by women	Frequency	Percentage (%)
Cocoyams	190	82.6
Cassava	176	76.5
Yams	185	80.4
Vegetables	195	84.7
Beans	202	87.8
Irish Potatoes	200	87
Maize	210	91.3

Source: Generated by Field Data.

Table 2.
Distribution of women by crops cultivated.

the active labour force in the food crop sector producing about 90% of all food crops in the country. Women grow food to feed the family, making subsistence food production one of the major agricultural activities of women [32, 37–39].

As shown on **Table 2**, women are responsible for the cultivation of food crops; as such, in the most part, food crop production is principally an activity of (rural) women, who combine the production, processing and marketing of food crops. They produce, process and market food crops such as vegetables, maize, beans, cassava, coco-yams, and irish-potatoes. As food crop producers, women continue to play a crucial role in ensuring family survival and safeguarding national food security. Women from urban centres also act as main food crop distributors, commonly referred to as 'buyam sellams', ensuring that food produced in rural areas are transported to urban towns and cities [4, 6, 14]. As buyam sellams, women are able to guarantee the availability of food to urban centres.

Despite women's predominant role in the agricultural sector as food crop producers, they have limited access to and control over production resources. The limitation of accessing and controlling production resources poses a significant difficulty not only for their productivity but for their ability to meet their family basic food needs. For example, women's ownership and control over land, labour and capital is very limited. Women lack ownership and inheritance rights to land, but maybe able to acquire use rights to cultivable land through male relatives. Consequently, with small landholdings comes the challenge of raising agricultural productivity for subsistence and generating surplus for the market.

6.2 Factors attracting land grabbing by local Investors in the Region

It is important to note that the study lays emphasis on the acquisition of large-scale land by local investors including elites in the local communities. In the North West Region, local investors have acquired great portions of land for the cultivation of agricultural plantations [10] including tea and palm plantations. One key common phenomenon that is also responsible for the acquisition of large-scale land in the region is the advent and presence of cattle grazers who are also responsible for grabbing large portions of cultivable land in the region. Land grabbing by cattle grazers, it is observed, is commonplace in all the Divisions in the Region and raises critical challenges with food crop production particularly with women, who are mainly responsible for subsistence food crop production. As confirmed by the interviews, a number of reasons were reported as responsible for the presence of local investors in the region. While the main local investor (Cameroon Tea Estate – CTE) acquires land for the purpose of cultivating tea, the major justification for interest in the region is associated with the weather and climate of parts of the region which is suitable for the cultivation of tea. The mountainous nature of the greater part of the region and the cold weather and climate in these areas particularly in Bui and Donga and Mantung Divisions is one of the major factors associated with the cultivation of tea in parts of these localities. In other divisions such as Momo, local investors are more interested in the production of palm oil given the suitability of the climate for the production of oil palm plantations. Some key informants also observed that one of the main reasons attracting small holding local investors is the availability of land in some of these villages. In relation to cattle grazing activities, land for grazing is by and large acquired through the authorization and permission of traditional rulers who are considered as custodians of village land. Meanwhile, the presence of and availability of cheap labour also significantly attracted local investors to acquire large portions of land for investment purposes and opportunities [17, 19, 21].

6.3 Community response to the phenomenon of LSLA by local investors

There has been diverse reactions from members of the community to the phenomenon as shown in other areas in the country; from youths, traditional rulers/chiefs, elites, human rights activists. There have been some responses from women's secret society as well. There is evidence that the land rights of women are threatened by the privatization and commoditization of land. Women are however responding to the limitations on land by accessing land through church membership and NGO-supported cooperatives [26, 27]. On the other hand, in some cases, women have also been reported to access land and improve their tenure security by way of resistance to some cases of land deals.

Some examples of these women's secret society include: the takembeng, fombuen and Anlu. I wish to reiterate here that these women's secret society in this part of the country can be seen as a base of power and agency for women. These secret societies have in the past been used as mechanisms for resistance and have been instrumental in safeguarding women's voice and access to resources especially access to farm land [40, 41].

The rush for land by local grabbers has taken over huge portions of farm land from farmers. The villagers especially women are of the opinion that the land grabs and rampant purchase of land has caused a reduction in cultivable land used by women for subsistence food crop production. Meanwhile, the villagers in general and women farmers in particular have been displaced from their farm lands making them to relocate further into the villages in search of new and free patches of cultivable land. This also meant that women farmers had to trek long distances or incur additional costs to travel to their farms. In addition, farmers are bound to change farms and pieces of land cultivated given that land becomes scarcer and getting a piece of cultivable land has become more difficult. Given that women farmers have relocated to new pieces of farmland, these are more likely to be abandoned land which is less productive with lower yields justifying one of the reasons for moving and changing plots of farmland. On the other hand, women's limited and lack of ownership rights to land also accounts for limitations and challenges of getting permanent farm land for the production of food crops either for subsistence or sale. According to evidence from the FGDs, some women farmers observe that they are bound to move further into the hinterlands looking for cultivable land in part to avoid land grabbers in case they are in need and in search for more land to purchase. This only confirms the difficulty for women to access farmland given the presence of agricultural plantations on the one hand and the associated effects of a fall and reductions in the production and availability of food crops for household consumption.

6.4 Land acquisition, local development and food security

As evidenced in the study, women are predominantly involved in subsistence food crop production while men concentrate on farming cash crops (coffee, oil palm), underlying women's frontline role in ensuring household subsistence and their contribution in the supply of household food needs and the maintenance of food availability. Thus, the contribution of women to food security cannot be undermined as women contribute more than 70% of the available food in the household and in the market. Women farmers confirmed the centrality of subsistence food crop production first as a major livelihood option and secondly in promoting household food security as they argued that even though men's cash crops are a source of income, providing for household subsistence and food needs is predominantly done by women as men are less concerned about food supply but concentrate on other children's needs such as paying school fees.

The findings speak to extant evidence about food security by showing that women's contributions to household food security is more than one might expect. We find that while women are at the forefront of ensuring household food security and care more about providing for and ensuring the availability of basic household food needs, men are more concerned with cash crop production. The evidence from the interviews observed that.

> *"...women in this community are the ones who struggle to provide food for the family; whether children eat or not is a major preoccupation of women and not men, who concentrate on their coffee and/or oil palm plantations, whose proceeds are often used to pay for the children's school fees".*

Despite promises of providing social amenities to the communities by local investors, this was scarcely fulfilled as confirmed by the findings. Local investors though reaping a lot of benefits from the communities where large scales of land were acquired did not provide any significant infrastructural development and investments in the communities. While some local investors promised to create employment, provide drinking water and electricity, farm to market roads, health centers, build schools and offer scholarships to the children of villagers, this was shelved and did not generate any meaningful development in these communities. The evidence from the survey and interviews show that local investors have made little or no contribution to improve on the health of villagers, neither have they provided farm to market roads and schools to the communities. Some of the villagers argue that *"roads constructed by local investors lead mainly to their plantations, while the hospitals provide free consultations only to staff and family members of staff"*. The villagers also concord that *"when we fall sick, we consult with the integrated health centres in our communities or buy medication from medicine stores; our children attend mainly government and private schools with very few one time and irregular scholarships provided by local investors"*. The lack of commitment by local investors in providing social amenities and infrastructural development to these communities is compounded by lack of accountability where according to [42, 43], the local investors are not made answerable to rural women and the villagers for breach of any development promises made prior to the acquisition. While noting that local investors created employment for villagers, it was argued that employment was mainly menial with very low wages and no social security. According to [17] some key functions and positions are often occupied by expatriates who meet the specific expertise and qualifications for the positions. Meanwhile, given women's low levels of education, they were by and large left out of the employment schemes of local investors.

The acquisition of large pieces of land for the cultivation of tea and palm oil in parts of the region, has taken huge amounts of village land with women affected disproportionately. Women lack ownership and inheritance rights though they can acquire use rights to cultivable land through male relatives [19, 21, 42, 43]. This limitation of increasing women's access to and control over land is exacerbated by the expansion of land grabbing phenomenon that has been observed in parts of Cameroon. This also places especially the rural women who scarcely enjoy security of tenure in a very precarious position as the least pressure on land as a result of high demand affects them as subsistence farmers. Women have experienced a reduction in cultivable land used for the production of food crops (vegetables, cassava, etc) and displacement from farmlands; thus, they have to walk for long distances very far into the villages in search of patches of available land for food crop production. In this regard, the land rush situation is more advantageous and beneficial to the investors leaving the communities and villages including the rural women (whose livelihood is highly dependent on land) in a worse off situation, notwithstanding the assumed benefits of socio-economic development and employment that such could bring [17, 43]. It is worth noting that land deals by and large miss the point that deals may to a large extent lead to landless peasants cum agricultural wage labourers [43] who may not be able to provide for basic household food needs, highlighting the consequences of deals on rural livelihoods and food security. The emerging evidence is that considering the meaning of food security as noted by [10], LSLA by local investors constraints the availability and access components of food security caused mainly by the displacement of women from cultivable farmlands. It also displaces the rural population in general and women in particular form their main livelihood options. From this perspective, local land grabs hinder and exacerbate women's limited access to land whose direct impact is a reduction in food crop production, a fall in women's farm related income, and a plausible shortage in the availability and access to basic household food needs.

7. Conclusion

In Cameroon, extensive evidence confirms the centrality of women in the agricultural sector in general and in subsistence food crop production in particular. They make up a majority of the active labour force in the food crop sector and as food crop producers and processors, women play a crucial role in ensuring family survival and safeguarding national food security. Although women play a crucial role in ensuring household food survival and food security, they lack access to and control over land, have limited ownership and inheritance rights to land with only use rights to land. This gender-based constraint serves as a critical limitation restricting women's access and control over resources, lesser livelihood options and their active participation in generating income for household sustenance. The expansion of large-scale land acquisition in the country only intensifies the scarcity of land, which is a major productive resource for women. As have others, this study argues that women bear the brunt of land deals particularly given limitations and challenges in relation to their control and ownership of land. While land acquisition by local investors displaces women from cultivable land, the land deals also deprive women of subsistence food crop production, a major source of livelihood for rural women. One is likely to conclude that the large-scale land deals by local investors including the promises of employment, social amenities and infrastructural development failed to trickledown to infrastructural and social development and more incomes for the local population with no positive effects on food security.

Author details

Christiana V. Abonge
Department of Women and Gender Studies, University of Buea, Cameroon

*Address all correspondence to: abonguec@yahoo.com

References

[1] Krieger J. Entrepreneurs and family well-being: Women's agricultural and trading strategies in Cameroon. In: Spring A, editor. Women Farmers and Commercial Ventures: Increasing Food Security in Developing Countries. Lynne Rienner: Boulder; 2000

[2] Endeley J. "Conceptualising women's empowerment in societies in Cameroon: How does money fit in"? Gender and Development. 2001;**9**(1):34-41

[3] Fonjong LN. Challenges and coping strategies of women food crop entrepreneurs in Fako division, Cameroon. Journal of International Women's Studies. 2004a;**5**(5):1-17

[4] Fonjong LN. Micro financial institutions as agents of improving women's economic empowerment in North Western Cameroon. Atlantis. 2002;**26**(2):120-130

[5] Fonchingong C. Structural adjustment, women and agriculture in Cameroon. Gender and Development. 1999;**7**(3):73-79

[6] Fonchingong C. Negotiating livelihoods beyond Beijing: The burden of women food vendors in the informal economy of Limbe, Cameroon. International Social Science Journal. 2005;**57**(184):243-253

[7] Amin AA. Introduction: Conceptual issues and overview. In: Amin AA, editor. Developing a Sustainable Economy in Cameroon. Dakar, Senegal: CODESRIA; 2008a. pp. 1-28

[8] Amin AA. Cameroon's economy and reforms. In: Amin AA, editor. Developing a Sustainable Economy in Cameroon. Dakar, Senegal: CODESRIA; 2008b. pp. 79-102

[9] Budlender D, Alma E. Women and Land: Securing Rights for Better Lives. Ottawa: IDRC; 2011. p. 201

[10] Fonjong LN, Gyapong AY. Plantations, women, and food security in Africa: Interrogating the investment pathway towards zero hunger in Cameroon and Ghana. World Development. 2021;**138**:105293

[11] Gobena M. The Effects of Large Scale Land Acquisition in Rural Ethiopia; The Case of Bako-Tibe Woreda. Uppsala: Swedish University of Agricultural Science; 2010

[12] Kachika T. Land Grabbing in Africa: A Review of the Impacts Anad the Possible Policy Responses. London: OXFAM; (n.d)

[13] Hall R, Paradza G. Pressures on Land in Sub-Saharan Africa: Social Differentiation and Societal Responses. ODI/DIE; 2012

[14] Childres M, Lastarria-Cornhiel S. Gender Issues in Land Policy and Administration: Overview. Washington, D.C: Word Bank; 2009

[15] Actionaid From under their feet: A think piece on the Gender Dimensions of Land Grab. 2012

[16] Lay J et al. Taking Stock of the Global Land Rush: Few Development Benefits, Many Human and Environmental Risks. Bern, Montpellier, Hamburg, Pretoria: Centre for Development and Environment; 2021

[17] Fonjong L, Sama-Lang IDN, Fombe L, Abonge C. Legalizing illegitimate large-scale land deals for commercial agriculture in Cameroon. Perspectives on Global Development

and Technology. 2018;**17**(1-2):116-136. DOI: 10.1163/15691497-12341470

[18] Fonjong, L; Sama-Lang, I. D. N; Fombe, L; Abonge, C. (2017) Large-Scale Land Acquisition and its Implications for Women's Land Rights in Cameroon. Ottawa, Canada. IDRC/CRDI

[19] RISD. Strengthening Women's Access to Land into IFAD Projects: The Rwandan Experience. Rwanda: IFAD/RISD; 2013

[20] Fonjong L, Sama-Lang IDN, Fombe L, Abonge C. Land governance and women's rights in large-scale land acquisitions in Cameroon. Development in Practice. 2016;**26**(4):420-430. DOI: 10.1080/09614524.2016.1162285

[21] Mutopo P. Gendered dimensions of land and rural livelihoods: The case of new settler farmer displacement at Nuanetsi Ranch, Mwanezi District, Zimbabwe. LDPI. PLAAS/ISS Working Paper No 8 2012

[22] Mutopo P, Chiweshe M. Large Scale Land Deals, Global Capital and the Politics of Livelihoods: Experiences of Women Small-holder Farmers in Chisumbanje, Zimbabwe. In: Paper presented at the International Conference on Global Land Grabbing II. October 17-19 2012. Ithaca: LDPI and Cornell University; 2012, 2012

[23] Adams M, Sianda S, Turner S. Land tenure reforms and rural livelihoods in Southern Africa. Natural resource perspective. No. 39. ODI. 1999

[24] Logo B, Bikie EH. Women and land in Cameroon: Questioning women's land status and claims for change. In: Women and Land in Africa: Culture, Religion and Realizing Women's Rights. London: Zed Books; 2003

[25] USAID. Country Profile: Property Rights and Resource Governance – Cameroon.

[26] UN Habitat. Profil Urbain National du Cameroun. 2010. http://www.unhabitat.org/pmss/listItemDetails.aspx

[27] African Development Bank. Cameroon: Diagnostic Study for Modernization of the Lands and Survey Sector. Yaounde, Cameroon: Country Regional Department Center; 2009

[28] Hobbs M. Cameroon Country Profile. Madison: Land Tenure Center, University of Wisconsin; 1998

[29] Nguiffo S, Djeukam R. Using the law as a tool to secure the land rights of indigenous communities in southern Cameroon. In: Cotula L, Mathieu P, editors. In Legal Empowerment in Practice: Using Legal Tools to Secure Land Rights in Africa. London: International Institute for Environment and Development (IIED); 2008

[30] Defo BK. Effects of socio-economic disadvantage and women's status on women's health in Cameroon. Social Science Medicine. 1997;**44**(7):1023-1042

[31] Endeley J. Structural Adjustment and the Cameroonian women's lifeline: 1986-1995. In: Risseauw C, Kamala G, editors. Negotiation and Social Space: A Gendered Analysis of Changing Kin and Security Networks in South Asia and Sub Saharan Africa. London: Sage Publications; 1998

[32] Pelican M. Mbororo Claims to Regional Citizenship and Minority Status in North-West Cameroon. Africa: The Journal of the International African Institute. 2008;**78**(4):540-560

[33] Scalise E. Indigenous Women's land rights: Case studies from Africa. State of the World Ministries and Indigenous Peoples. 2012:53-59

[34] World Bank. African Development Indicators. New York: World Bank; 2001a

[35] UNDP. Human Resource Development: Cultural Liberties in today's Diverse World. New York: UNDP; 2004

[36] Endeley JB. Strategies and progress for women in agricultural sector in Africa. WID Process in Berlin (West). DSE 1985

[37] Fonjong LN. Changing fortunes of government policies and its implications on the application of agricultural innovations in Cameroon. Nordic Journal of African Studies. 2004b;**13**(1):13-29

[38] Logo B, Patrice and Elise-Henrietta Bikie. Women and Land in Cameroon: Questioning women's Land Status and Claims for Change. Atlanta: University's Women and Land Studies; 1998

[39] IDRC. Promoting Accountability and Legitimacy in Relation to Large Scale Land Acquisitions. 2013

[40] Polack E, Cotula L, Cote M. Legal Empowerment and Pathways to Accountability in the Global Rush for Africa's Land. London: IDRC; 2013

[41] Fonjong LN, Sama-Lang IF, Fombe LF. Implications of customary practices on gender discrimination in land ownership in Cameroon. Ethics and Social Welfare. 2012;**6**(3):260-274. DOI: 10.1080/17496535.2012.704637

[42] Fonjong LN et al. Land Tenure Practices and Women's Rights to Land in Anglophone Cameroon: Implications for Sustainable Development. IDRC. Lambert Academic Publishing; 2013

[43] Ngwa NE. Introduction to land and rural women in Cameroon. In: Fonjong LN, editor. Issues on Women's Land Rights in Cameroon. Limbe, Cameroon. 2012. pp. 1-17

Section 4

Natural Disasters and Food Security

Chapter 7

Review of Food Access in the Shade of Critical Times

Abrar Almalki, Balakrishna Gokaraju and Raymond C. Tesiero

Abstract

Food security is a condition where people at all times have physical, economic, and social access to sufficient, safe, and nutritious food. Even though the world produces enough food for all humans, we have people with low access to food due to several socioeconomic variables, and this lack of food access affects their lives and health in return. On regular days, physical access to food outlets is limited by several variables and that creates categories in food access areas. In extreme weather events, the accessibility becomes more limited due to the impact on mobility, and interruption to electricity or transportation systems. These weather events increase due to climate change, and the future is highly unpredictable. The cumulative effect of climate change and pandemics increases the risk of complicated circumstances for food security. COVID-19 pandemic is considered the largest epidemic, and the onset of these large-scale future epidemics is predicted to be inevitable. This chapter will describe the extent of food access categories in Guilford county of North Carolina pre-COVID-19, and the influence of climate change and the COVID-19 pandemic on the food-access distribution.

Keywords: food security, food access, health issues, chronic diseases, climate change, natural disasters, pandemic, COVID-19

1. Introduction

It is ongoing research on how the food system influences safe and healthy diets and in turn helps toward maintaining a sustainable and resilient environment. The food system includes production, processing, packaging, distribution, marketing, purchasing, consumption, and waste food [1]. The food system addresses food security, food safety, and food quality. Food safety is food protection from food-borne illnesses, covers, handling, preparation, and storage of food. The quality of food provides nutritional benefits to the people. Food security includes food access during all times and under any condition. Food access becomes critical under severe climate events, natural disasters, pandemics, and political stress. These events influence access negatively which results in food insecurity.

2. Food security

According to The United Nation Food and Agriculture Organization (FAO), the definition of food security includes four dimensions: (i) the availability of sufficient quantity and quality food; (ii) the access by people to nutritious food; (iii) supporting well-being systems with freshwater and health care; and (iv) stable food sources [2]. Stability is the ability to access food and the availability of food at all times. Food security is a current global concern for the next 50 years and more. Food security is unequally distributed and areas such as sub-Saharan Africa and South Asia are suffering the most from food insecurity. Food securities also include food availability, affordability, food quality, and food access. Food availability also is dependent on the production, distribution, amount, and types available for purchase at stores. Food affordability is the economic cost affordability of an item. Food quality is acceptable nutrition. People's food access is influenced by their income levels, access to resources, physical environment, social environment, cost of food, their government, and trade policies.

Food security is affected by policy, environment, socioeconomic, and culture. The socioeconomic factors mean low-income neighborhoods, low access to public transportation, and a low number of food sources. Increasing the food security in a region or a county would need a collaboration of policies, investments, human resources development, agriculture research, rural infrastructure, water resources, farm infrastructure, and natural resources management. Over the 7 billion population, we have 2 billion in food-insecure because they meet one or more of the FAO's definitions of food security [3]. The population would reach between 8.3 and 10.9 billion people by 2050 and that makes food security challenges due to the current starving and undernourished rates. This hunger and lack of access to food vary in different tier-level countries, such as developed countries it is 5%, the developing countries reached 13%, African countries reached 20%, and 13% in Asian countries. Concerning the increasing population, the world will need to double or increase by 70% of the food production by the year 2050. Ten percent of the world is undernourished, i.e. 821 million in 2018. This 10% is distributed in three continents, where Asia is ranked as the highest with 514 million, Africa with 256 million, and South America and the Caribbean with 42.5 million.

Vulnerable areas to food security are also defined as "Hot Spots" [4]:

1. Vulnerable areas affected by war are facing difficulties in logistics and distribution during pandemics and normal conditions.

2. Vulnerable areas are affected by extreme weather events and pests that affect crop yield.

3. Vulnerable poor communities were 821 million people already facing food insecurity before the pandemic of COVID-19.

4. Vulnerability to food cost. These countries face major currency depreciation and that decreases their ability to import food items.

3. Food access

People's ability to access food physically and financially has an important impact on their lives. Measuring food access can be done by several methods and on several levels. At the individual level in the United States, the National Food Stamp Survey in 1996–1997 documents people's answers about their food outlets' distance to their homes [5]. More surveys were developed in 2011 as a food security supplement survey asking if they had enough food in their homes [5]. Another measure that investigates the indirect impact is the distance to food outlets by the Economic Research Service (ERS) 2012 [5]. Further measure at the individual level is the Retail Food Environment Index (RFEI). It is a survey method about people's access to food outlets and their options [5]. Another method of measuring access to food is by the area category location of food outlets and related variables in a neighborhood [5]. In the area category location method, the access to food can be measured by the distance to food outlets, income, poverty rates, and transportation access [5]. These factors categorize tracts or districts from the poorest in healthy options to the most nutritious area. These areas are as follows:

Food desert: This phrase was firstly used in Scotland in the 1990s to define low access to food outlets due to low income [6]. After that, researchers around the world adopted the definition and applied more classifications. It is defined as areas with more than one mile to grocery stores. People living in food desert areas or tracts have limited healthy options with higher prices and limited access to public transportation [7]. The area's residents rely on small stores more than supermarkets because of location availability [7]. These small stores tend to have smaller quantities and higher prices, and fewer fresh options [7].

Food swamp: These areas or tracts have more unhealthy food outlets than healthy food options [8]. These neighborhoods are more likely to be minority and low-income [9]. Several studies proved that fast-food restaurants are concentrated in racial-ethnic minorities' neighborhoods [10].

Food oases: These areas have people with high socioeconomic status and an abundance of healthy food outlets such as supermarkets, grocery stores, farmers' markets, and community gardens [11]. This food access condition is the ideal goal for food access development where people have more options for healthy food.

The influence of food access on people's health and behavior is investigated, and it varies based on income, urbanism, population density, and people's ethnicity. Furthermore, low access to food causes several negative effects on people's health such as delayed childhood development, diet-related diseases, and malnutrition or hypernutrition. These food access areas show different records and impacts on people's lives and health [12]. Factors such as car access and income are the most critical variables in people's access to food [13]. A study in Australia found that financial aid or allowance improved people's access to food [14]. For that, government financial aid and food credits can support people's access to healthy food. People's access to transportation such as busses stations or personal cars improves their nutritional choices. In addition, walking to food outlets is not feasible in different seasons, summer limits food access due to the heat during the day, and winter limits food access due to the extremely cold climate [15]. Lack of access to adequate food is a key factor in some diseases such as obesity, diabetes, hypertension, cancer, and high mortality rates [6]. A study conducted in the United States proved that food swamps can predict obesity more than food deserts, and its prediction becomes more accurate when it is combined with income inequality and low transportation access [9]. Moreover, a study conducted in Australia proved the same hypothesis that access to transportation is more affected by people's access to food than their location in food desert areas [16].

Inadequate distribution of food outlets in urban areas influences people's well-being negatively. People's health and wellness are proven to show a negative impact on food access in several studies in the United States and around the world. Several studies found that people with more access to fast food also indicate having higher body mass [17]. A study conducted in Philadelphia, a city ranked as the highest in food scarcity, in the surrounding area by 2 miles of fast-food restaurants, found out that the low access to fresh food sources is associated with a higher rate of health issues such as diabetes, heart diseases, and cancer [18]. A study in California illustrates obesity variation to correlate with food access, but the variation in obesity could be because of people's characteristics and diets [19]. A study compared people in the food desert and found out that people in the food desert showed more chronic kidney disease and hypertension [20]. Moreover, a recent study is conducted by our team in Guilford County in North Carolina on the correlation between accessing food and health issues [12]. Their results illustrate that food deserts show higher mortality rates and higher rates of health issues such as obesity, high cholesterol, and high blood pressure [12]. More studies on the correlation between food distribution and people's health proved a strong correlation with illnesses such as tuberculosis [21].

Furthermore, it is documented that there is a connection between food deserts, low income, low access to transportation, and diabetes [22]. Food and physical activity have a significant impact on predicting children's obesity [23].

More variables or factors influence food access such as ethnicity. In the United States, low access to food areas is in low-income and colored communities [24]. A study illustrated the investigation of people's ethnicity and their food access and choices [24]. The survey results show variation in food access based on ethnicity [24]: less healthy food was consumed more by African Americans; Latino communities have better access to healthy food; and immigrants consumed more fruit, rice, and beans [25].

3.1 Food access and climate change

The earth's climate is changing, and this change has influenced people and the environment. Influences such as population growth, migration, need for water, shelter, and livelihood have a major influence on changing the earth's equilibrium. That leads to changes in the environment such as deforestation, biodiversity loss, habitat loss, desertification, and scarcity [1]. Most of these changes are made by humans to urbanize their cities. Regarding man-made activities, there is evidence that burning fossil fuels and land use changes made cumulative effects on these events. As defined by Paul Crutzen and Eugene Stoermer, the era when human activities affect the earth, and the environment is the Anthropocene Epoch [1]. The most obvious sign of this era is climate change. These climate changes are caused by human activities on natural elements such as the increase of greenhouse gas (GHG) emissions, carbon dioxide, and methane. Carbon dioxide has been increasing from 284 ppm in 1832 to 397 ppm in 2013 [3]. Methane is a strong gas and increased from 340 ppm in 1980 to more than 420 ppm in 2022. It is about the changes in the atmospheric temperature that influences more changes. This greenhouse gas (GHG) emission increases and changes to land use are the main cause of climate change [15]. Earth temperature has been increasing since the nineteenth century. Since the 1850s, the temperature increased by 0.8°C according to documented temperature increases in oceans and lands [3]. The temperature is expected to increase by 1.8°C which is higher than in previous centuries. More recently, a report by the National Center For Environmental Information, NOAA, illustrated the temperature for more than 143 years from 1880 to 2022 [26]. The monitoring of temperature in

DOI: http://dx.doi.org/10.5772/intechopen.107041

March over the years found that this year's temperature is 0.95°C (1.71°F) above the twenty-first century average (March 2022).

Climate change effects are direct and indirect and could affect some places more severely than others [27]. Climate change influences food security in six aspects [3]: 1—The impact is greater in high hunger countries; 2—it will increase undernutrition and malnutrition; 3—it will increase food inequalities on all scales global and local; 4—people in vulnerable conditions to extreme weather events will suffer more because of the severity of climate change; 5—the climate change effects will continue affecting the earth in the next 20–30 years because of the past carbon emissions; and 6—severe weather events will increase in severity and its future is unpredictable.

As climate change is becoming more severe sources such as clean water, air quality, low pathogen exposure, and the ability to raise, gather, harvest, and gather crops, animals, seafood, and wild food will be threatened [28]. These threats to food availability will negatively affect all human nutritional status [28]. Climate change affects food security by affecting crop production, storage, and transportation [1]. It influences the food supply chain, demand, and access on a local scale.

Since the 1990s, decreases in food security, commodity prices increase, and the decrease in per capita cultivated are documented [29]. Climate changes affect the availability of food items specifically fruit and vegetables, and it is predicted to be between 5 million to 175 million by 2080 [2]. For instance, 30% of farmers in developing countries are food-insecure [30]. In tropical and worming environments, climate change would severely increase rainfall, and it will cause pests and disease to the crops and livestock [31]. Regions such as African Sahara are influenced by biophysical, political, and socioeconomic variables, which make them more vulnerable to climate change [32]. One of the climate change impacts is the loss of cropland, and the sub-Sahara in Africa is expected to lose more than 10–20% of its crop and West Africa by 20–50% by 2050 [32]. These changes and droughts influence rain pattern changes, which has been estimated that East Africa is likely to become wetter and Southeast Africa to become drier [32]. The complexity of environmental tension with lack of food access raises the stress on vulnerable communities.

Based on climate forecasting models for the next 100 years, the land temperature will be increasing and result in lowering the crop yield in dry regions and increasing the short-term yield in high-latitude regions. These changes have a consequence on providing food, water, and shelter for humans. These affect climate changes such as temperature increases and changes in rainfall. Climate changes may influence hunger, but they could be controlled by feeding people during crises and investigations into new agricultural techniques that increase yields. Moreover, monitoring and predicting systems can be used by the government for early warning [30]. Insurance cover crops depend on remote sensing to cover the needed crops in case of drought years [30]. Investment in modified seed has more ability to survive [30]. More local policies protect local production [30]. The temperature increases are likely to impact crops negatively in the next two decades and affect corn wheat, rice, and other primary crops [30]. One of the expected effects of climate change in Mali is a hunger increase from 34% to 44% [33].

3.1.1 Food access and natural disasters

Climate change influences severe events such as natural disasters, scarcity, low air quality, unequal food distribution, and conflict over natural sources. Natural disasters hit the world differently and in some regions the effects are severe. Each natural disaster impacts several environmental elements.

Drought: It is the decrease of rainfall rates more than usual per season or year. Drought can be in several forms such as meteorology drought known as precipitation shortage, agriculture drought as soil moisture, and socioeconomic drought as influenced by other drought types [34]. The increases in drought events influence food security negatively because of water scarcity and lower agriculture yield, which affects their availability in food outlets and results in a lack of food access. Moreover, it influences the availability of food, which increases its prices because of scarcity [34]. The drought in 2007–2008 combined with increased energy prices caused a food crisis, where food prices increased more than double [35]. In the Middle East, the hydrological drought peak was in 2009 as 50% of the area faced extreme drought. Meteorological drought was documented in the early 1970s, and 2000s, and 60% drought in the period between 2008 and 2012. These drought events affected food security, especially in 2010.

Hurricane: During severe rain and hurricane events, life sectors got affected at different levels. The hurricane may affect electricity and streets. The electricity damage stops all activities dependent on it such as food refrigeration. Damages to trees and falling on roads may stop people from accessing essential needs such as food outlets. Regardless of the effect on all stores, large food outlets such as supermarkets seem to be more resilient. Natural disasters such as hurricanes can affect food sources and that results in low food availability. For instance, in Puerto Rico, 69% of farmers were experiencing food insecurity after hurricane Maria in 2017, and 38% of them were experiencing food insecurity for 3 months or more [36].

Flooding: It is when water overflows or soaks in an area that is not a river or a lake. A natural disaster like flooding has several impacts on the food system. Floods impact agriculture negatively and that affects the food available for purchase. Crops were damaged by the water overflow. In Bangladesh, the flood damaged 12% of their rice yield in 2007 [37]. In Pakistan, the flood of 2010 resulted in a crop loss of approximately 19% [37]. Some countries' experiences with floods and other natural disasters encouraged them to make an impact such as what governments of the Philippines, Malaysia, and ideas planned [36]. In the United States, severe events such as floods and hurricanes are documented between 2004 and 2014, and the country recorded 24 natural disasters with effect on 1.90 million people. These events affect people's access by damaging infrastructure which limits people's access to food [38].

3.1.2 Adapting to climate change in the food sector

Despite the fact that climate change effects are expected to increase in the next 40 years, food production needs to be increased by 50% to face the land shrinking and to provide necessary food [26]. Although we are adopting the current changes, preparedness for unexpected future changes is also highly critical. Some countries started adapting to climate change effects. For instance, Ghana adapted to the drought and rainfall changes by planning and installing early-maturing, drought-resistant crops, and applying agrochemicals [32]. In Burkina Faso, West Africa, they planned dry season crops and installed micro water harvesting systems. Several considerations are to be taken in dealing with climate change to protect food [32].

Several recommendations related to policies are identified as follows [4]:

1. Merging food security and sustainable agriculture into global and national policies.

2. Increasing the global investment in sustainable agriculture and food system in the next decade.

3. Reducing the greenhouse gas emission and negative impacts while shifting agriculture production to more sustainable productions.

4. Programs and policies targeting vulnerable communities to climate change and food insecurity.

5. Reshaping people's access to food and their consumption patterns for better nutrition, healthier, and sustainable eating pattern.

6. Reducing waste in all food systems: infrastructure, farming practice, processing, distribution, and household habits.

7. Establish a system including human and ecological sectors that share, integrate, and comprehensively all information.

3.2 Food access during pandemic

Throughout history, mankind has been exposed to several pandemics and epidemics that affected their health and life. In history, humanity faced several epidemics such as the Spanish Flu, Asian Flu, Hong Kong Flu, HIV/AIDS, SARS, Ebola, and Swine Flu [4]. In the last centuries, humans faced Bubonic Plague in the 1300s, smallpox in the 1500s, Spanish Influenza in 1918, and HIV/AIDS in the 1980s. Recently, humans faced SARS in 2002, Bird Flu in 2003, Swine Flu in 2009, MERS in 2012, Ebola in 2012, and Zika in 2015 [39].

A pandemic affects all human activities, and one of them is agriculture and farming production. Limiting people working in these production areas results in food shortages in stores. The limited quantities in food stores and outlets will increase food scarcity. The food items scarcity would increase their prices and add more burden to access. Furthermore, the pandemic spread would require people to be quarantined to protect themselves and that would limit their visits to stores.

A recent pandemic is COVID-19 which is caused by a respiratory disease [18]. The first COViD-19 cases were discovered in Wuhan, China, on January 18, 2020 [40]. After the spread of cases in several countries, the World Health Organization declared COVID-19 as a global pandemic on March 11, 2020 [41]. Based on sequencing technology on the COVID-19 virus, it was found 96.2% of bats as a possible source of SARA-CoV2 [42]. When the pandemic started, the United Nations World Food Program warned the world that by the end of 2020 approximately 265 million people would suffer food insecurity, which is more than 135 million people in food insecurity conditions before the pandemic.

Countries such as the United States announced restrictions on traveling and interactions [43]. Firstly, face masks were mandated mostly in every country and social distancing was highly encouraged [43]. Moreover, negative COVID-19 tests were required in emergencies, 24 hours before travel, to limit the spread in destination countries [43]. Schools were shut down and students adapted to online education for the first time in history. Some countries enforced curfews to limit people's interactions and the spread of the virus. The goal of these restrictions was to contain the pandemic spread, but they had a negative psychological influence on communities.

At the beginning of the pandemic, India had only 36 COVID-19 cases and 7 death cases. On April 5, 2020, the cases were 3577 and 83 deaths but rose rapidly to 508,953 cases and 15,685 deaths by Jun 28, 2020. The government took actions to control the spread by restricting traveling and imposing a lockdown effectively from March 24, 2020, to May 24, 2020, at different levels. In the United States, the first case was documented on January 20, 2020 [44].

In India, the pandemic changed people's consumption patterns [45]. In rural areas, people's income was affected by the lockdown, and people in villages consumed lower-quality food during the pandemic [45]. In Australia, according to a survey study, households' food security decreased because of the pandemic, and that affected people's habits and cooking.

In the beginning phase of the pandemic, people panicked and started buying excess food and household goods. That led to limitations on people's access to sufficient food and changed their food habits and cooking. The lockdown restriction impacted the supply chain through a labor shortage and that affected food availability and access. The pandemic reduced 25% of labor. Moreover, the pandemic affected food availability in poor countries and with low national security. Consumer shopping patterns have changed due to the pandemic increases. The pandemic encouraged businesses and farmers to change their business models and use online ordering, pickup, and delivery.

The FAO admitted the huge loss of food and agriculture by the pandemic. Dealing with the pandemic influence, the FAO advised countries to focus on food requirements for vulnerable communities, supporting social safety programs and supporting small local farmers. The food index documents food prices monthly and during the pandemic in 2020 prices of cereals, oilseeds, dairy products, meat, and sugar had the lowest point since December 2018. For example, palm oil production in Indonesia and Malaysia was affected by the pandemic labor shortage which resulted in restrictions on exporting palm oil from January 2022 through March of the same year [46]. Moreover, according to the FAO, the pandemic affected small farmers and fishermen, who had declined in their supplies and that declined their ability to cover their finances.

The COVID-19 pandemic affected the supply chain strongly and affected food security, especially for vulnerable communities, and they have less food storage to last during the pandemic. There are 10 countries with 103.3 million people suffering from a food crisis [47]. In countries like India, the demand increased for long shelf life vegetables such as onion and potato but decreased for short shelf life vegetables such as fruits and grapes. Developed countries were affected, but their dependence on technologies helped them subsidize the labor work, and developing countries were affected severely because of their independence from human labor. The fear of contracting the disease is strong on people's mental health as a secondary effect of the pandemic, and it was proven that mental disorders are higher in communities with food insecurity [48]. The restricted lockdown affected people's mental health which increased depression and that increased people's food consumption and panic buying [48]. Furthermore, vulnerable people to contract the virus are people facing chronic health conditions such as cardiovascular disease, diabetes, chronic respiratory disease, hypertension, and cancer.

This pandemic caused an international food emergency. In Europe, the pandemic increased the food demand by more than 50% with the lockdown [49]. In Europe, food supply stores were banned from exporting without warning to deal with the pandemic uncertainty [49]. In Europe, food consumption increased due to the lockdowns and remote working [17]. Spain took similar restrictions with the world and declared a state of emergency on March 14, 2020 [17]. In Spain and Italy, the pandemic restrictions encouraged people to adopt healthier diets [17]. In Spain, people were experiencing food shortages in stores by 64%. Then pandemic people started stockpiling nonperishable food items and buying extra items every trip to avoid the shortage [17]. In turn, it caused a change to buy food from local farms and organic sources [17]. In Spain, younger people less than 40 years of age increased their food consumption more than people over 40 and 50 years old during the pandemic [17]. In Germany, Denmark, and Slovenia, people consumed longer shelf life food and less fresh food. In the United States, because of the pandemic, people changed their food source and where they eat. Since the beginning of the pandemic to January 2021, the United States had more than 21 million infection cases and 350,000 deaths, which count as 20% of the global total cases. Since the pandemic onset, the unemployment rate increased to 7.9% in September 2020, a level not seen since the Great Depression. By December, 14% of adults were not having enough food across the United States.

Before the pandemic in 2017, three Arab countries were importing 100% of their cereals, namely Qatar, Kuwait, and Bahrain, and the United Arab Emirates at 99% [50]. Moreover, seven Arab countries import approximately 90% of their grains, namely Jordan, Amman, Palestine, Yemen, Saudi Arabia, Libya, and Lebanon [50]. Algeria is also at high risk of food crisis depending on importing 79% of its grains [50]. Countries like Tunisia, Mauritania, and Iraq import more than 55% of their grains. Egypt imports 44% of its [50]. The least dependent Arab countries are importing 40–33% of their grain in Morocco and Sudan [50]. In Algeria, products like semolina, flour, dry grains, and pasta became unavailable because of the demand increase and that led to prices increase, and the government took the lead to secure food sources and control monopoly and speculation [50].

The responses to the current pandemic may change the environment, and it is an opportunity to create a multidisciplinary solution for human livelihood and future generations. The Avian Influenza (H5N1) pandemic in 1997 in Hong Kong led them to create "One Health" to predict, control, and minimize future outbreaks [51]. The One Health is a multidisciplinary work in local, national, and global stakeholders [51]. Some considerations to lower the pandemic effect on different regions such as Africa: 1—develop the health system capacity by the public fund in Africa; 2—financial aid needs to be distributed among individuals, entrepreneurs, and corporations, so they contribute to lowering the negative impact of the pandemic; 3—during the pandemic and lockdowns, employees should be given incentives and; 4—central banks in African countries need to support domestic banks to save credit and liquidity. One of the consequences of COVID-19 in Africa is putting approximately 29 million people under the extreme poverty line as expected by the United Nations Economic Commission for Africa (ECA).

Because of the land use change and more human interaction with animals and the environment, more transmittable diseases are emerging. Changes in climate and people activities and land use changes cause habitat disruption and lead to pathogen movement [47]. Emerging infectious diseases (MIDs) are increasing with climate change and make a dangerous nexus. These diseases are increasing the food cost and lowering production which limits access to food. By 2030, poor counties are expected to import their food by 2%. Support from the framework is needed for these accruing matters. A framework called Document Assess Monitor Act (DAMA) is applied to deal with these issues.

1. Document: by identifying pathogens based on the current information available and their occurrence location before. Then, document the pathogen and its incident by time and place and consequences.

2. Assess: to detect related microbes, place them under the monitor. This phase includes two steps: A) phylogenetic triage and studying the pathogen. When it discovers a chance of disease spread, it will be reported to the authorities, and B) analyzing these concerning pathogens and connecting them to old diseases to estimate a prediction of spread location.

3. Monitoring: by resampling these potential threats regularly and investigating vulnerable areas to the spread.

4. Act: after the invention of the current situation, the management plan forms to formulate an action plan to act on a specific situation.

Following are the important issues recommended in the pandemic era:

1. Consumers are changing their food habits to a healthier diet. Foods rich in vitamins and nutrition support the immune system such as vitamin C. Foods rich in vitamins such as carrot and spinach are known for their ability to fight susceptibility to infections.

2. Safety from the virus spread on products. Food items can carry the SARS-CoV-1 virus and people after touching it and touching their face can catch the virus. For that, carrying bags of food items needs to be followed by intensive hand washing. Studies documented that the COVID-19 virus can live up to 2 years in a frozen state.

3. Food security due to the lockdown. Food shortages during a pandemic affect people's health and increase hunger. For that, the government should maintain emergency plans to keep all food supplies going.

4. Food sustainability. Fighting hunger by supporting food security is in the United Nation for sustainable development. Sustainable food sources contribute to healthy lives and well-being.

Future recommended actions to deal with future pandemics are given as follows [52]:

1. Additional support to the sustainable development goals, and assistance to local and small producer communities and backyard gardens in low middle-income countries.

2. More engagement and understanding between consumers and producers to support the food system's resilience to pandemics.

3. Identifying sequences and legislation to deal with it.

4. Identifying future risks and setting up risk management plans.

5. More support and funding for scientists to do research development and training in food systems.

6. Support the perspective of "One Health" to encourage all countries to work together on mitigating pandemics' effects on people's health around the world.

3.3 Food access during war

Food access becomes extremely critical when a country is under political stress. War is one of the events that create ripple effects on different countries outside the war scene. It demoralizes humanity that it could happen to any ordinary person in this situation. When a war happens in a country, its residents and business lose their ability to export and import food, which leads to low access to food for people in the war country, its neighbors, and importers. However, a war in one country can affect the whole world by affecting its food production as is the current case with Ukraine and the Russian war.

According to the world health organization, the Syria war that started on March 15, 2011, is the worst humanitarian crisis in the twenty-first century [53]. However, even before the war, people of Syria were having difficulties accessing freshwater 93% in urban areas and 86% in rural areas [53]. This war impacted people's lives, education, and health. The war complicated the access to food and water. It is documented that 90% of people of Syrians were having difficulties accessing food [53]. Moreover, one in seven children is not fed regularly [53]. Documented studies illustrate that 32% of children had signs of malnutrition [53]. Furthermore, approximately 62% of infants did not have access to the best substitute feeding [53].

The war between Eritrea and Ethiopia from 1998 to 2000 caused a food crisis in both countries [54]. These two countries are among the vulnerable countries to food security [54]. Before the war started, the world health organization warned that more than 16 million people are starving because of the drought, 50% in Ethiopia and the rest in Kenya, Somalia, and Uganda [54]. Moreover, 40% of Ethiopian farm households were in chronic food insecurity [54]. After the war started, food items prices increased, and that affected people's access [54]. Moreover, food was weaponized by soldiers, and farmers were stopped from harvesting their crops which led to food unavailability for people to access [55].

The recent food crisis is going on in the whole world in 2022. Because Russia invaded Ukraine on February 24, 2022, the country was not able to export its yield to its regular customers [56]. This limitation on exporting affected food security, fertilizer prices, and energy globally [57]. The FAO report on food cruises because of the war in 2022. In March 2022, food prices were higher than average by 29.8% compared to the previous year in 2021. Countries started hoarding food in fear of inflation and food crises. The effect of the war on food is greater in countries facing food insecurity, especially with the rise of people suffering insecurity from 768 million in 2020 to 869 million in 2022. Countries that export most of their wheat from Russia and Ukraine such as Egypt, Lebanon, and Yemen are affected severely by the food shortage [34]. Moreover, countries were facing food insecurity and drought such as Ethiopia, Kenya, and Somalia [34]. Even before the war started, Ukraine and other countries were having high prices. Ukraine wheat prices were historically high because of energy and oil prices, drought, and the effect of the COVID-19 pandemic.

During the Russian and Ukraine war, both countries announced they are unable to export grains. Their announcement led to some panic buying and country hoarding.

Countries started to ban exporting their crops to protect their markets and make them available to their residents at controlled prices [58, 59]. For instance, India limited its wheat to deal with the loss of 15–20% of its crop because of the heat wave [34]. India is the second largest wheat producer, and Indonesia which produces 60% of the world's palm oil restricted limitations on these food goods [56]. Argentina banned beef, soybean, and soybean meal [57]. Russia banned sunflower seeds, wheat flour, and pasta [57]. Ghana banned the exportation of maize, rice, and soybean [56]. Algeria banned pasta, wheat, and vegetable oil [56]. Egypt banned vegetable oil, corn, wheat, wheat flour, and pasta [57]. Serbia banned wheat, corn, and wheat flour [57]. Kuwait banned grains and vegetable oil [57]. Other countries followed suit such as India, and they stopped their grain imports. Countries like Kazakhstan announced their limitation on flour and wheat on April 1, 2022, for 6 months [57]. Wheat producer countries in the black sea region have 30% of the wheat in storage to feed the world [34]. The loss of wheat from Ukraine could be filed by countries including India, Argentina, Australia, and Canada, but shipping prices are high due to energy prices [34].

4. Conclusion

Food security and equality are about the availability and safe access to nutritional food. These characteristics of food security influence people's health and their chronic conditions. Unequal access to healthy food impacts people's health negatively. People's access to food may be limited by their lack of transportation access and low income. Improving these conditions would improve people's access to health and fresh food. Food insecurity is influenced by several conditions such as environmental changes, pandemics, and wars. These conditions show a severe impact on vulnerable countries' food security. The combination of more than one condition increases the risk of food insecurity. The combination of the COVID-19 pandemic with the Russian invasion of Ukraine increased the food crisis globally and food importation. The impact of these food access areas can be improved over time. Several considerations can be applied in future to improve the food access conditions, and they are government support by adding more public transportation stops, supporting low-income communities through policies and financial aid, increase young people's nutritional awareness in schools, increasing the percentage of fresh fruit and vegetables in convenient stores, improving the healthy food options in food swamps.

Acknowledgements

We thank the North Carolina Department of Environmental Quality for sponsoring the faculty compensation and supporting this study using the ARPA-NCDEQ funds, 2021. We extend thanks to College of Engineering, Department of Computational Data Science and Engineering for administrative support.

Conflict of interest

The authors declare no conflict of interest.

Author details

Abrar Almalki[1*], Balakrishna Gokaraju[2] and Raymond C. Tesiero[3]

1 Geography and Geographic Information System Department, King AbdulAziz University, Jeddah, Saudi Arabia

2 Visualization and Computation Advancing Research Center (ViCAR), North Carolina A&T State University, Greensboro, NC, USA

3 Civil, Architectural, and Environmental Engineering Department, North Carolina A&T State University, Greensboro, NC, USA

*Address all correspondence to: aaalmalki7@kau.edu.sa

References

[1] Fanzo J, Bellows AL, Spiker ML, Thorne-Lyman AL, Bloem MW. The importance of food systems and the environment for nutrition. The American Journal of Clinical Nutrition. 2021;**113**(1):7-16

[2] Schmidhuber J, Tubiello FN. Global food security under climate change. Proceedings of the National Academy of Sciences. 2007;**104**(50):19703-19708

[3] Wheeler T, Von Braun J. Climate change impacts on global food security. Science. 2013;**341**(6145):508-513

[4] Mouloudj K, Bouarar AC, Fechit H. The impact of COVID-19 pandemic on food security. Les cahiers du CREAD. 2020;**36**(3):159-184

[5] Ver Ploeg M, Dutko P, Breneman V. Measuring food access and food deserts for policy purposes. Applied Economic Perspectives and Policy. 2015;**37**(2):205-225

[6] Abel KC, Faust KM. Modeling complex human systems: An adaptable framework of urban food deserts. Sustainable Cities and Society. 2020;**52**:101795

[7] Jin H, Lu Y. Evaluating consumer nutrition environment in food deserts and food swamps. International Journal of Environmental Research and Public Health. 2021;**18**(5):2675

[8] Statement of Policy, Healthy Food Access. NACCHO: Washington, DC, USA. Available from: https://www.naccho.org/uploads/downloadable-resources/13-04-Healthy-Food-Access.pdf

[9] Cooksey-Stowers K, Schwartz MB, Brownell KD. Food swamps predict obesity rates better than food deserts in the United States. International Journal of Environmental Research and Public Health. 2017;**14**(11):1366

[10] Fleischhacker SE, Evenson KR, Rodriguez DA, Ammerman AS. A systematic review of fast food access studies. Obesity Reviews. 2011;**12**(5):e460-e471

[11] Charreire H, Casey R, Salze P, Simon C, Chaix B, Banos A, et al. Measuring the food environment using geographical information systems: A methodological review. Public Health Nutrition. 2010;**13**(11):1773-1785

[12] Almalki A, Gokaraju B, Mehta N, Doss DA. Geospatial and machine learning regression techniques for analyzing food access impact on health issues in sustainable communities. ISPRS International Journal of Geo-Information. 2021;**10**(11):745

[13] Sisiopiku VP, Barbour N. Use of GIS spatial analysis to identify food deserts in the state of Alabama. Athens Journal of Health. 2014;**1**(2):91-104. DOI: 10.30958/ajh.1-2-2

[14] Kent K, Murray S, Penrose B, Auckland S, Godrich S, Lester E, et al. Food insecure households faced greater challenges putting healthy food on the table during the COVID-19 pandemic in Australia. Appetite. 2022;**169**:105815

[15] Widener MJ, Shannon J. When are food deserts? Integrating time into research on food accessibility. Health & Place. 2014;**30**:1-3

[16] Coveney J, O'Dwyer LA. Effects of mobility and location on food access. Health & Place. 2009;**15**(1):45-55

[17] Li S, Kallas Z, Rahmani D, Gil JM. Trends in food preferences and sustainable behavior during the COVID-19 lockdown: Evidence from Spanish consumers. Food. 2021;**10**(8):1898

[18] Sohrabi C, Alsafi Z, O'Neill N, Khan M, Kerwan A, Al-Jabir A, et al. World Health Organization declares global emergency: A review of the 2019 novel coronavirus (COVID-19). International Journal of Surgery. 2020;**76**:71-76

[19] Truong K, Fernandes M, An R, Shier V, Sturm R. Measuring the physical food environment and its relationship with obesity: Evidence from California. Public Health. 2010;**124**(2):115

[20] Suarez JJ, Isakova T, Anderson CA, Boulware LE, Wolf M, Scialla JJ. Food access, chronic kidney disease, and hypertension in the US. American Journal of Preventive Medicine. 2015;**49**(6):912-920

[21] Gehlen M, Nicola MRC, Costa ERD, Cabral VK, de Quadros ELL, Chaves CO, et al. Geospatial intelligence and health analitycs: Its ap-plication and utility in a city with high tuberculosis incidence in Brazil. Journal of Infection and Public Health. 2019;**12**(5):681-689

[22] The Demographic Statistical Atlas of the United States - Statistical Atlas. 2022. Available from: https://statisticalatlas.com/about. [Accessed: July 16, 2022]

[23] Ohri-Vachaspati P, DeLia D, DeWeese RS, Crespo NC, Todd M, Yedidia MJ. The relative contribution of layers of the social ecological model to childhood obesity. Public Health Nutrition. 2015;**18**(11):2055-2066

[24] Gordon C, Purciel-Hill M, Ghai NR, Kaufman L, Graham R, Van Wye G. Measuring food deserts in New York City's low-income neighborhoods. Health & Place. 2011;**17**(2):696-700. DOI: 10.1016/j.healthplace.2010.12.012

[25] Misra AK. Climate change and challenges of water and food security. International Journal of Sustainable Built Environment. 2014;**3**(1):153-165

[26] NACCHO Healthy Living and Prevention Workgroup. STATEMENT OF POLICY Healthy Food Access, (Approved by NACCHO Board of Directors N0. 13-04), Washington DC. 2013

[27] Gitz V, Meybeck A, Lipper L, Young CD, Braatz S. Climate change and food security: Risks and responses. Food and Agriculture Organization of the United Nations (FAO). 2016;**110**:2-4

[28] Chakraborty S, Newton AC. Climate change, plant diseases and food security: An overview. Plant Pathology. 2011;**60**(1):2-14

[29] Reddy VR, Singh SK, Anbumozhi V. Food Supply Chain Disruption Due to Natural Disasters: Entities, Risks, and Strategies for Resilience. Jakarta, Indonesia: ERIA Discussion Paper Series; 2016

[30] Brown ME, Funk CC. Food security under climate change. Science. 2008;**319**(5863):580-581

[31] Gregory PJ, Johnson SN, Newton AC, Ingram JS. Integrating pests and pathogens into the climate change/food security debate. Journal of Experimental Botany. 2009;**60**(10):2827-2838

[32] Connolly-Boutin L, Smit B. Climate change, food security, and livelihoods in sub-Saharan Africa.

Regional Environmental Change. 2016;**16**(2):385-399

[33] Butt TA, McCarl BA, Angerer J, Dyke PT, Stuth JW. The economic and food security implications of climate change in Mali. Climatic Change. 2005;**68**(3):355-378

[34] Hameed M, Ahmadalipour A, Moradkhani H. Drought and food security in the middle east: An analytical framework. Agricultural and Forest Meteorology. 2020;**281**:107816

[35] McDonough S. Countries are limiting food exports. It may make global hunger worse. 2022. [online] Vox. Available from: https://www.vox.com/future-perfect/23140267/countries-are-limiting-food-exports-it-may-make-global-hunger-worse. [Accessed: June 07, 2022]

[36] Rodríguez-Cruz LA, Álvarez-Berríos N, Niles MT. Social-ecological interactions in a disaster context: Puerto Rican farmer households' food security after hurricane Maria. Environmental Research Letters. 2022;**17**(4):044057

[37] Pacetti T, Caporali E, Rulli MC. Floods and food security: A method to estimate the effect of inundation on crops availability. Advances in Water Resources. 2017;**110**:494-504

[38] Lassa JA, Teng P, Caballero-Anthony M, Shrestha M. Revisiting emergency food reserve policy and practice under disaster and extreme climate events. International Journal of Disaster Risk Science. 2019;**10**(1):1-13

[39] Chin A, Cui X, Gillson L, Nelson D, Taylor MP, Vanacker V, et al. Anthropocene in an age of pandemics. Anthropocene. 2020;**30**:100247

[40] Xu H, Yan C, Fu Q, Xiao K, Yu Y, Han D, et al. Possible environmental effects on the spread of COVID-19 in China. Science. Total Environ. 2020;**731**:139211

[41] Khan N, Siddiqui BN, Khan N, Ullah N, Wali A, Khan IU, et al. 2. Drastic impacts of COVID-19 on food, agriculture and economy. Pure and Applied Biology (PAB). 2020;**10**(1):62-68

[42] Galanakis CM. The food systems in the era of the coronavirus (COVID-19) pandemic crisis. Food. 2020;**9**(4):523

[43] 2022. Available from: https://www.cdc.gov/quarantine/travel-restrictions.html. [Accessed: July 16, 2022]

[44] 2022. Available from: https://covid.cdc.gov/covid-data-tracker/#datatracker-home. [Accessed: July 16, 2022]

[45] Suresh V, Fishman R, von Lieres JS, Rao BR. Impact of the COVID-19 lockdown on the economic situation and food security of rural households in India. Journal of Agribusiness in Developing and Emerging Economies. 2022;**12**(3):491-509

[46] Fao.org. FAO Food price Index| World Food Situation | Food and Agriculture Organization of the United Nations. 2022. Available from: https://www.fao.org/worldfoodsituation/foodpricesindex/en/. [Accessed: June 07, 2022]

[47] Trivellone V, Hoberg EP, Boeger WA, Brooks DR. Food security and emerging infectious disease: Risk assessment and risk management. Royal Society Open Science. 2022;**9**(2):211687

[48] Clay LA, Rogus S. Impact of employment, essential work, and risk factors on food access during the

COVID-19 pandemic in New York state. International Journal of Environmental Research and Public Health. 2021;**18**(4):1451

[49] Farcas AC, Galanakis CM, Socaciu C, Pop OL, Tibulca D, Paucean A, et al. Food security during the pandemic and the importance of the bioeconomy in the new era. Sustainability. 2020;**13**(1):150

[50] Beddington JR, Asaduzzaman M, Bremauntz FA, Clark ME, Guillou M, Jahn MM, et al. Achieving Food Security in the Face of Climate Change. France: Commission on Sustainable Agriculture and Climate Change; 2012

[51] Sims LD, Peiris M. One health: The Hong Kong experience with avian influenza. In: One Health: The Human-Animal-Environment Interfaces in Emerging Infectious Diseases. Berlin, Heidelberg: Springer; 2012. pp. 281-298

[52] Mardones FO, Rich KM, Boden LA, Moreno-Switt AI, Caipo ML, Zimin-Veselkoff N, et al. The COVID-19 pandemic and global food security. Frontiers in veterinary. Science. 2020;**928**

[53] Elsafti A, van Berlaer G, Al Safadi M, Debacker M, Buyl R, Redwan A, et al. Children in the Syrian civil war: The familial, educational, and public health impact of ongoing violence. Disaster Medicine and Public Health Preparedness. 2016;**10**(6):874-882. DOI: 10.1017/dmp.2016.165

[54] White P. War and food security in Eritrea and Ethiopia, 1998-2000. Disasters. 2005;**29**:S92-S113. DOI: 10.1111/j.0361-3666.2005.00286.x

[55] Stanton M. Ethiopian War Weaponized Access To Food Causing A Man-Made Famine. 2022. Available from: https://theowp.org/ethiopian-war-weaponized-access-to-food-causing-a-man-made-famine/#:~:text=In%20Tigray%2C%20access%20to%20food%20is%20being%20weaponized,medical%20aid%20to%20turn%20around%20and%20leave%20Tigray [Accessed: June 21, 2022]

[56] Duggan L. Countries are hoarding food supplies As prices spike, hunger crisis looms. 2022. Dailycaller.com. Available from: https://dailycaller.com/2022/05/25/countries-hoarding-food-supplies-prices-spike-hunger-ukraine-russia/. [Accessed: June 07, 2022]

[57] Donley A. List of countries restricting grain exports growing. 2022. World-grain.com. Available from: https://www.world-grain.com/articles/16917-list-of-countries-restricting-grain-exports-growing#:~:text=List%20of%20countries%20restricting%20grain%20exports%20growing.%201,%-28grains%2C%20vegetable%20oil%2C%20through%20Dec.%2031%2C%202022%29.%20. [Accessed: June 07, 2022]

[58] Listiyorini E, Raghu A Rahadiana R. Bloomberg. 2022. Bloomberg.com. Available from: https://www.bloomberg.com/news/articles/2022-04-27/indonesia-to-halt-crude-palm-oil-exports-in-expanded-ban. [Accessed: June 07, 2022]

[59] Nationthailand. India may curb rice exports to control price in domestic market. 2022. Available from: https://www.nationthailand.com/international/40016012. [Accessed: June 07, 2022]